Topics in Applied Physics
Volume 86

Available Online

Topics in Applied Physics is part of the Springer LINK service. For all customers with standing orders for Topics in Applied Physics we offer the full text in electronic form via LINK free of charge. Please contact your librarian who can receive a password for free access to the full articles by registration at:

http://link.springer.de/orders/index.htm

If you do not have a standing order you can nevertheless browse through the table of contents of the volumes and the abstracts of each article at:

http://link.springer.de/series/tap/

There you will also find more information about the series.

Springer
Berlin
Heidelberg
New York
Hong Kong
London
Milan
Paris
Tokyo

Topics in Applied Physics

Topics in Applied Physics is a well-established series of review books, each of which presents a comprehensive survey of a selected topic within the broad area of applied physics. Edited and written by leading research scientists in the field concerned, each volume contains review contributions covering the various aspects of the topic. Together these provide an overview of the state of the art in the respective field, extending from an introduction to the subject right up to the frontiers of contemporary research.

Topics in Applied Physics is addressed to all scientists at universities and in industry who wish to obtain an overview and to keep abreast of advances in applied physics. The series also provides easy but comprehensive access to the fields for newcomers starting research.

Contributions are specially commissioned. The Managing Editors are open to any suggestions for topics coming from the community of applied physicists no matter what the field and encourage prospective editors to approach them with ideas.

See also: http://www.springer.de/phys/series/TAP

P. Boffi D. Piccinin M.C. Ubaldi (Eds.)

Infrared Holography for Optical Communications

Techniques, Materials, and Devices

With 90 Figures
and 10 Tables

Springer

Dr. Pierpaolo Boffi
Dr. Davide Piccinin
Dr. Maria C. Ubaldi

Commutazione Ottica Milano (CoreCom)
Consorzio Ricerche Elaborazione
Via Ampère 30
20131 Milan
Italy

Email:
boffi@corecom.it
piccinin@corecom.it
ubaldi@corecom.it

Library of Congress Cataloging-in-Publication Data.

Infrared holography for optical communications: techniques, materials, and devices
P. Boffi, D. Piccinin, M. C. Ubaldi (eds.) p.cm. – (Topics in applied physics, ISSN 0303-4216)
Includes bibliographical references and index. ISBN 3540433147 (alk. Paper)
1. Holography. 2. Infrared imaging. 3. Optical storage devices.
I. Boffi, P. (Pierpaolo), 1966- II. Piccinin, D. (Davide), 1968- III. Ubaldi, M. C. (Maria C.), 1970- IV. Series.
TA1540.I54 2002 621.36'75–dc21 2002022167

Physics and Astronomy Classification Scheme (PACS): 42.40.E, 42.70.L, 42.79.S

ISSN print edition: 0303-4216
ISSN electronic edition: 1437-0859
ISBN 3-540-43314-7 Springer-Verlag Berlin Heidelberg New York

Springer-Verlag Berlin Heidelberg New York
a member of BertelsmannSpringer Science+Business Media GmbH

http://www.springer.de

Typesetting: DA-TEX Gerd Blumenstein, Leipzig
Cover design: *design & production* GmbH, Heidelberg

Printed on acid-free paper SPIN: 10868840 56/3141/mf - 5 4 3 2 1 0

Preface

Umbrella: *... used as a protection against the sun in China, Egypt, and elsewhere in remote antiquity. ... its usefulness was not rediscovered until the late 16th cent., when it was introduced as the parapluie (Fr.=against the rain). (The Columbia Encyclopedia, Sixth Edition. 2001).*

What's umbrella got to do with holography? It is just a question of evolution in use... Consider a fascinating and well-established technology as holography (our "umbrella") employed since many years in the conventional field of data storage (the use against the "sun"). Such as umbrella was successfully introduced as repair against the rain after centuries of a different and well established use, nowadays holography can be rediscovered in the optical fiber communication network!

Joking aside, fiber communications based on transmission of optical signals at wavelengths set in the low-loss fiber attenuation windows (in near-infrared range) make possible to deliver new broadband services to the end users: this use implies the necessity to re-design the optical equipment inside the network in order to replace electronics and remove its constraints. We believe that a compelling combination of both the experimented maturity of holographic technology and new clever solutions and architectures based on holographic trickeries can give holography the required firepower to penetrate inside the strategic multibillion-dollars business market of optical communication components. This book will not present a general overview on holography, but will focus on the state-of-the-art information about holography exploitation in the field of optical communications. Selected contributions providing a comprehensive coverage of the matter as far as related techniques, materials and devices are contained. After a preliminary review regarding up-to-date optical data storage, main recording techniques useful to transfer the powerful advantages of holography to near-infrared range are presented. Basic material aspects are then shown, also taking into account the aim to build integrated-optics devices. Hologram erasure in time has been always the Achilles' heel of holographic devices, causing the slowing down of their commercial debut. Long-lifetime reliability granted by fixing techniques is hence considered. Finally, some experimental implementations are described: they perform fundamental functionalities inside the fiber network and represent preliminary examples of IR holography-based components for optical communications.

It is our hope that the joint effort of many highly qualified contributors will help the reader to gain a better insight into "edge" research on IR holography and will suggest him new smart device ideas, which can make a heavy impact on future optical systems and network solutions.

Our heartfelt thanks to all the contributors for bringing this book into being.

Milano, Italy
April 2002

Pierpaolo Boffi
Davide Piccinin
Maria Chiara Ubaldi

Contents

Review of Optical Data Storage

Daniel Day[1], Min Gu[1], and Andrew Smallridge[2]

[1] Centre for Micro-Photonics, School of Biophysical Sciences and Electrical Engineering, Swinburne University of Technology
PO Box 218 Hawthorn, Victoria 3122, Australia
{dday,mgu}@groupwise.swin.edu.au

[2] School of Life Sciences and Technology, Victoria University of Technology
PO Box 14428 MCMC, Melbourne, Victoria 8001, Australia

Abstract. As the computer industry grows, so will the requirements for data storage. Magnetic memory has been the most stable method in terms of capacity and recording/reading speed. However, we have reached the point where a substantial increase in the capacity cannot be produced without increasing the size of the system. When compact discs (CDs) were introduced in the 1980s they revolutionized the concept of data storage. Since their inception, the capacity requirements have far exceeded what is available on a compact disc, and they are now following the same path as magnetic memories. Following this trend, it could be assumed that digital versatile discs or digital video discs (DVDs) have a limited lifetime as a storage medium. In fact it has been noted that the maximum capacity of DVDs will be reached in 3–5 years. The question then is what comes next. This chapter aims to illustrate the technology involved in current optical storage methods as well as to introduce several new concepts of optical storage. It is envisaged that a storage system based on either solid immersion lens, holography or three-dimensional bit recording could be the way of the future.

The development of optical technology has revolutionized the way we communicate between people or between computers. As society continues to require better tools to communicate more data at higher rates, so does it require the ability to store larger amounts of information. Since the invention of the first computer there has always existed the need for some form of information storage system other than printed hardcopies. One of the first of such systems was computer ribbon; although somewhat awkward it freed the user from having to input the required information at the beginning of every session. At the time this was one of the greatest advances in computer technology. Several other data storage systems have followed over the years, and each time there has been a limit to the amount of information that could be stored.

1 Introduction to Optical Data Storage

In the last few years we have witnessed the use of audio cassettes, floppy disks (magnetic media), compact discs (CDs), digital versatile discs or digital video discs (DVDs) (optical mediums) and now the slow emergence of a hybrid technology magneto-optical disc (MO). Throughout this period there have also been improvements in the way that the information has been encoded and

P. Boffi, P. Piccinin, M. C. Ubaldi (Eds.): Infrared Holography for Optical Communications,
Topics Appl. Phys. **86**, 1–21 (2003)

transmitted. New techniques for compressing data has led to another format of storing information that is fast becoming popular for audio tracks. Moving Picture Experts Group (MPEG) files are compressed to such an extent that the capacity of a compact disc is now available on a small memory chip, using the latest compression MP3 (3rd generation).

However, all of the above systems have or will reach a finite limit beyond which they cannot increase the storage capacity of the devices. In the case of MPEG systems, they can only increase the capacity of the memory chip by increasing the size of the chip or decreasing the size of the components used in designing the chip. Increasing the size of the chip will work up to a point beyond which the system becomes impractical. The size of the components is bound by a limit below which the components become physically impossible, and economically unfeasible to manufacture. For optical systems (CDs and DVDs) the finite limit is imposed by the wave nature of light: the size of a minimum resolvable point is usually no less than half of the wavelength of the light used to image it. Therefore, the amount of information that can be stored on an optical disk is limited by the wavelength of the light used to record or read the information.

A new technology that has emerged and is based on optical recording, but is not restricted by the wave nature of light, is near-field optical data storage. In the near-field region, diffraction is not a dominant effect and so the size of the recorded bit is limited by the optical system used. Typically there are two devices used for recording and reading in the near-field region: a fibre probe or a solid immersion lens (SIL). Both systems are capable of producing a recorded bit 10 times smaller than that in CDs or DVDs, but again both the fibre probe and SIL systems are limited to recording one layer of information near the surface of the recording material.

All of the above-mentioned recording systems could be classified as two-dimensional recording systems, where they only record one layer near the surface of the material. The recording materials that are used in CDs and DVDs are manufactured to be 1.2 mm thick, in which case the recording systems use 0.01% of the volume of the material. In terms of the storage capacity per device, such two-dimensional systems are very inefficient. To effectively utilize the other 99.99% of the volume we need to investigate three-dimensional recording and reading systems.

In the rest of this chapter, we will provide an introduction to the different types of optical data storage systems. A review of compact discs, digital video discs and magneto-optical discs is covered, as well as the areas with emerging technologies, holographic and solid immersion lens.

1.1 Compact Discs/Digital Video Discs

In 1983 a collaboration between Philips and Sony saw the introduction of compact discs into the consumer market [1]. Within three years CDs were selling at over one million per year. At the time the capacity of a CD was no more than what was currently available on magnetic cassette tapes; however, it introduced the ability to record and replay audio tracks in digital quality sound.

Recording information in digital format reduces any interference that would corrupt the quality of the information, but as a drawback it requires a tremendous amount of storage space; one second of digital audio requires over one million bits.

The optical setup in the reading system is based upon the interference of the light reflected from the pit and the land. The discs are fabricated such that the light reflected from the land has traveled half a wavelength more than that from the pits and therefore destructively interferes producing no reflection from the pits.

The standard design for optical systems incorporates a laser diode operating at an appropriate wavelength for reading either a CD or a DVD. A diffraction grating follows the laser diode and has the effect of producing a main peak and two side-lobes which are used in the tracking mechanism. The three peaks then pass through a polarizing beam splitter which transmits only parallel-polarized light, followed by a collimator. A quarter-waveplate is used to convert the light to circular polarization before being focused down onto the CD/DVD. If the light strikes land, then it is reflected back through the objective and converted by the quarter-waveplate back into linearly polarized light; however, this time with vertical polarization. The polarizing beam splitter then reflects the vertical polarized light through a focusing lens and a cylindrical lens onto a quadrant detector. The cylindrical lens is used in the auto-focusing mechanism. A schematic diagram of the optical system used in CDs and DVDs is shown in Fig. 1.

Within ten years the capacity of a CD fell behind what was required of storage devices, and the DVD emerged onto the market. Early versions of the DVD were able to store 4.7 Gigabytes (1 byte is equal to 8 bits) of information, almost 7.5 times more information than a CD. Current predictions have the DVD limited to 25 Gigabytes per disc, if double-layer, double-sided technology is used, within the next five years [2].

Table 1 illustrates the changes that were made to the optical system for CDs and DVDs to increase the capacity of the system.

By using a shorter-wavelength laser and a higher numerical aperture (NA) objective the DVD optical system is able to produce a smaller focused spot, therefore reducing the minimum pit length and track pitch. This reduction produces an increase in the storage capacity per layer.

As well as increasing the density per layer in DVDs, further development has been conducted into producing double-layer, double-sided DVDs, which would increase the capacity by 2 and 4 times respectively.

The technology involved in DVDs is such that further increases beyond 25 Gbytes in storage capacity are unlikely. Due to the multi-layered material structuring involved in one recording layer, the signal from the second layer

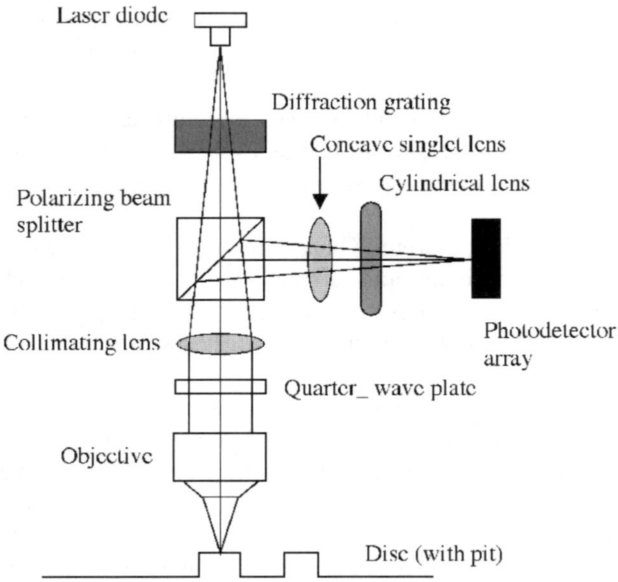

Fig. 1. The schematic diagram of the optical system used in CDs and DVDs

Table 1. Comparison of the optical parameters between CDs and DVDs

Parameters	CD	DVD
Diameter	120 mm	120 mm
Thickness	1.2 mm	1.2 mm
Laser Wavelength	780 nm	640 nm
Numerical Aperture	0.45	0.60
Minimum Pit Length	0.834 μm	0.40 μm
Track Pitch	1.6 μm	0.74 μm
Data Capacity (per layer)	0.68 Gbytes	4.7 Gbytes
Layers	1	1,2,4

is significantly degraded, and therefore further layers on the same side are impossible. The high tolerances on the thickness of the layers in DVDs increases the cost of manufacturing a disc. It has been estimated that the cost of one recordable DVD will equal the cost of twenty CDs. At this point in time the capacity of twenty CDs is greater than that of one DVD.

1.2 Magneto-Optical Discs

Recordable magneto-optical discs are quite different to CDs in the way that data is recorded and retrieved. On a conventional CD, microscopic pits reflect light from a laser beam. Their presence or absence makes up the digital code that is converted into a music signal. On a MO disc it is the polarity of a magnetic field that makes up the digital code. During recording, a laser beam heats a minute portion of the disc while a recording head on the opposite side writes the code by changing the polarity of the magnetic field. Then for playback the laser reads the disc by detecting differences in light reflected by the coded magnetic layer. Magneto-optical discs are immune to adverse magnetic influences (unlike standard cassettes) as they need to be heated to around 180 °C for the polarity to be altered. Figure 2 shows the optical and magnetic setup and the recording mechanism of a MO system.

This method of data storage has been used successfully in computer applications for some time and is extremely reliable and durable. In fact, it allows a disc to be re-recorded up to a million times with no loss of quality. The longevity of MO memory far surpasses any tape format, and has been estimated by Sony at well over thirty years with no loss of quality [1].

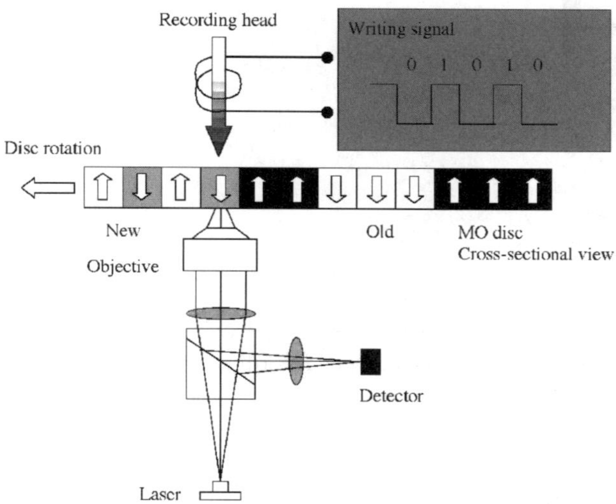

Fig. 2. Recording mechanism in magneto-optical discs

1.3 Solid Immersion Lens

All of the different systems described above work in the far-field region where the maximum resolution and therefore the data density are defined by the wave nature of light. By introducing a specially designed high-refractive index medium between the objective and the recording medium, the effective numerical aperture of the system is increased. Figure 3 shows the typical configuration of a solid immersion lens recording system.

The SIL lens is normally designed to be a hemisphere or super-hemisphere, with a refractive index greater than 1.9. Materials such as GaP with a refractive index of 3.3 have been used [3]. An appropriate combination will result in a recording system with an effective numerical aperture as high as 1.9. Such a high numerical aperture allows the SIL lens to work using total internal reflection.

Solid immersion technology has been demonstrated with both MO [4] and phase change [3] recording media. However, a limitation of the system is the tolerance on the 100 nm air gap between the SIL and the recording medium. Changes in the size of the gap dramatically affect the signal contrast of the readout system [5]. The presence of dust on the recording surface will easily destroy the performance of the recording system.

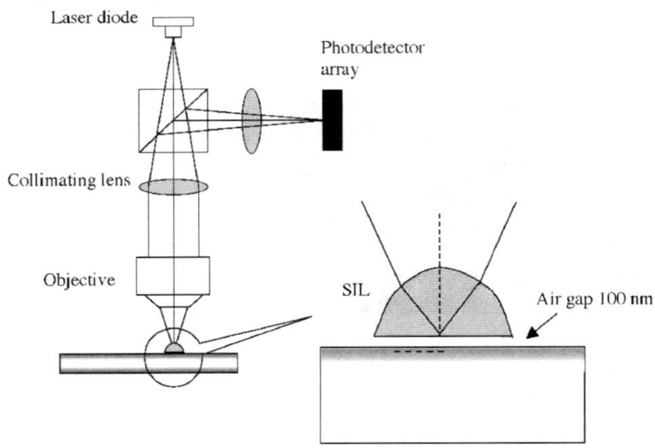

Fig. 3. Schematic diagram of a solid immersion lens recording system

1.4 Holographic Storage

The concept of holography is accredited to Dennis Gabor who was attempting to improve the image quality of electron microscopy in 1947. Since the 1970s holography has been applied to optical data storage [6]. While the density of this method of recording is expected to reach the limit of Tbit/cm^3, the data transfer rates are far superior to that of conventional storage systems. Due to its ability to record and read out a whole plane at a time, a transfer rate somewhere between 1 and 100 Gbits/s is predicted [7]. A comparison of the achievable recording densities between holographic and multi-layered bit recording by Tanaka and Kawata in 1996 [8] summarized that for holographic storage to reach Tbits/cm^3 it has to employ angle multiplexing.

Two techniques used by holographic storage to record multiple pages of information within the same region are angle and wavelength multiplexing. By slightly changing the angle or wavelength of the reference beam multiple holograms can be recorded on top of each other without interference. An advantage of holographic storage is that the information can be randomly accessed if the recording conditions (i.e. reference angle) are known.

There are several very similar techniques for holographic recording and the method described below is just one example. For two-photon holographic recording the sample is placed at the spatial and temporal intersection of two beams (Fig. 4). The first beam (probe beam) is focused onto a thin sheet of light in the recording medium. The second beam (pump beam) is passed through a spatial light modulator (SLM) after being expanded and collimated. The SLM is computer controlled and can impart a desired recording pattern onto the collimated pump beam. After passing through the SLM the pump beam is then recollimated and imaged onto the plane illuminated by the probe beam.

A problem with most holographic data storage systems is that quite often the reading system used erases the recorded information. Researchers have been working on different methods to solve this problem such as thermal fixing or periodic rewriting of the recorded information.

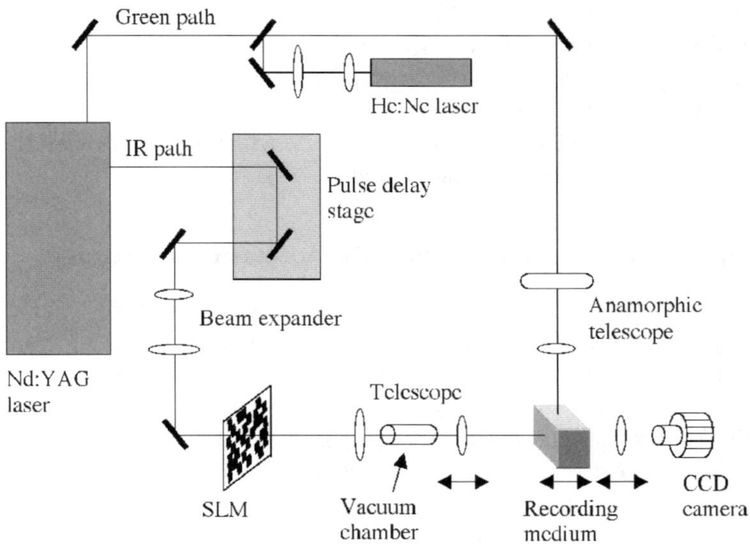

Fig. 4. Schematic diagram of a holographic recording and reading system

1.5 Three-Dimensional Bit Optical Storage

Three-dimensional bit data storage [9] is another technique whereby information can be recorded within the volume of a recording medium. The pits described in CDs and DVDs are the result of the stamping process that produces the discs, whereas the bits created in three-dimensional storage are a chemical/physical change in the material (not necessarily a recessed region). The materials and methods of three-dimensional bit optical data storage are summarized in Sect. 2

2 Three-Dimensional Bit Optical Data Storage

This section reviews the advances made in three-dimensional bit optical data storage. The materials used are of particular importance as they determine the type of recording and reading methods that can be used.

According to the diffraction theory of light in a far-field region, the light distribution in the focus spot has a certain size, which is primarily dictated by the wavelength of the light and the numerical aperture of the objective. The shorter the wavelength and the higher the numerical aperture, the smaller is the resulting diffraction pattern in the focal region of the objective. It is this property which limits the capacity of optical data storage systems. Current optical data storage systems only record information within a two-dimensional plane near the surface of the material, using approximately 0.01% of the available volume in a CD or a DVD. If the third spatial dimension is used to record information, there is an instant increase in storage capacity without increasing the volume of the storage medium. As optically thick recording media are currently being used, a future system would benefit from being able to record in an identical volume medium, which would provide the basis for a next-generation backwards-compatible system.

2.1 Principle of Three-Dimensional Bit Optical Data Storage

In three-dimensional bit data storage, information is stored in three dimensions by recording a layer of bits (information) in the transverse (x–y) plane near the surface, and then successive layers are recorded at different depths into the material (Fig. 5). By focusing a laser beam into specific materials, different types of physical and chemical changes are created. The number of layers that can be recorded within the volume of the material is dependent on the axial resolution of the recording and reading methods. Higher axial resolution will allow the distance between layers to be reduced and therefore increase the storage capacity.

However, when a recording beam is focused into a volume medium, scattering caused by the medium occurs; the shorter the wavelength the stronger the scattering process [10]. As a result, the energy carried by the recording

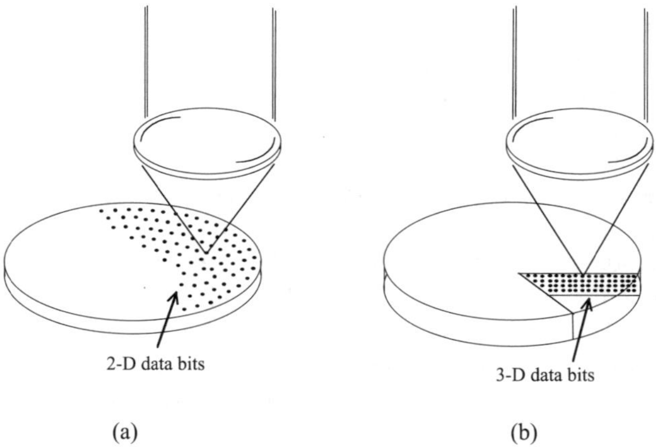

Fig. 5. Schematic diagram for (**a**) 2-D and (**b**) 3-D optical data storage

beam cannot be efficiently transferred into a deep position in the recording medium [11]. To overcome this problem, a two-photon excitation process has been adopted [12,13].

2.2 Single-Photon Versus Two-Photon Excitation

As indicated in Fig. 6 the principle difference between single- and two-photon excitation is the absorption of one and two incident photons, respectively. In single-photon excitation, the absorption of a photon (typically from the ultraviolet (UV) to visible region) promotes an electron from the ground state to an excited state. The energy of the absorbed photon is given by

$$E = h\nu \tag{1}$$

where h is Planck's constant and ν is the frequency of the incident photon. The process of two-photon excitation requires that two photons, each having energy of $E/2$, be absorbed simultaneously to excite an electron from the

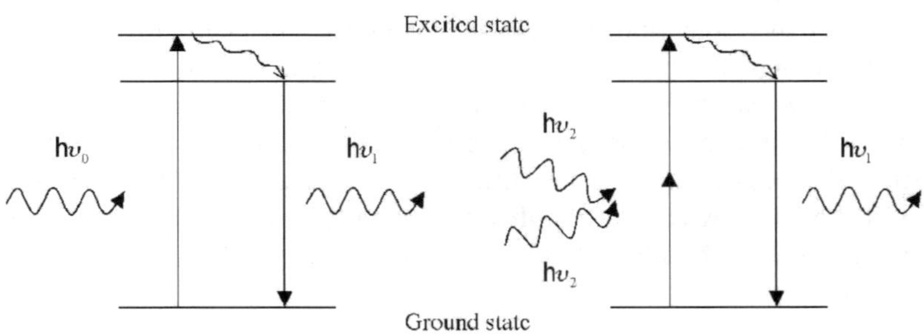

Fig. 6. Energy level diagram for (**a**) single-photon and (**b**) two-photon excited fluorescence

ground state to an excited state [13]. For this transition to happen there is the requirement that both photons are spatially and temporally coincident, which is a third-order nonlinear effect [14].

The energy required to promote an electron from the ground state to an excited state typically corresponds to the energy of a photon with a wavelength in the UV-visible region of the electromagnetic spectrum. Most compounds manufactured to increase the photosensitivity of the recording materials have absorption bands in this region. This therefore requires the use of a laser and optics designed to work at the shorter wavelengths. A disadvantage with this is the high UV absorption in the glass used for the optics.

In two-photon excitation, the nonlinear relationship between the incident light intensity and excitation means that the probability of excitation is significantly lower than that for single-photon excitation [14]. Therefore, an ultra-short pulsed laser with a pulse width of a few hundreds of femtoseconds is used to increase the excitation efficiency. It should be noted that the use of continuous-wave illumination for two-photon excitation in biological tissue has been demonstrated. Continuous-wave two-photon excitation would allow for a small high-powered laser diode to be used as the laser source, making a recording system more practical.

The probability of two-photon excitation is lower than that for single-photon excitation because two-photon excitation is a third-order nonlinear process [14]. Due to the quadratic dependence of the excitation on the illumination intensity, the excitation is confined to a small volume within the focal region of the objective. The linear absorption probability of single-photon excitation means that excitation occurs almost along the entire illumination path. The highly localised excitation of two-photon excitation is a direct result of the nonlinear absorption. The localised excitation results in a property known as optical sectioning. Optical sectioning provides the ability to record a layer of information above or below a previous layer without an overlap of information, otherwise known as crosstalk.

Another advantage of two-photon excitation is the use of a near-infrared wavelength to excite the materials in the UV-visible region. According to Mie scattering theory [10], the shorter the wavelength, the larger the scattering cross-section. Therefore, when focusing deep into the medium the photons are likely to scatter more than for focusing near the surface. This becomes a significant problem when the bit and layer spacing is reduced to near the diffraction limit. *Denk* et al. [13] for the first time, reported on the use of two-photon excitation in conjuction with laser scanning fluorescence microscopy. At the time this was deemed a breakthrough in imaging, as it was possible to excite UV dyes with a near-infrared wavelength while achieving high resolution with less probability of photobleaching living cells.

2.3 Photopolymerization Effect

Strickler and *Webb* [12] were the first to demonstrate the ability to produce high-density optical data storage using two-photon excitation. They achieved a density as high as 0.3 Tbits/cm^3, with a bit spacing of 1 μm and a layer spacing of 3 μm using a photopolymerizable solution.

In photopolymerization, a gel solution consisting of a monomer and a photoinitiator are combined in a cell. Upon illumination the photoinitiator produces free radicals that start the polymerization of the monomer. Using two-photon excitation, the polymerization can be confined to within the excitation region of the focus spot. It is ideal to irradiate the sample with UV light before recording, so as to gel the sample to prevent distortion of the recorded planes from shrinkage or flow.

As the sample polymerizes, a change in material density occurs at the recorded bit. This change corresponds to a change in refractive index of 0.8% for Cibatool XR5081 [12], a change from 1.541 for the monomer to 1.554 for the polymer.

Such a large change in the refractive index for the recorded bit can then be read using a phase-sensitive microscope. Differential interference contrast (DIC) microscopy can be used to produce a phase/intensity map of the recorded pattern, thereby effectively reading the pattern of recorded bits.

Table 2 covers the different materials and equipment that have been used to record information within three dimensions using photopolymerization.

It should be noted that the ability to fabricate structures with resolution that is close to the diffraction limit could be useful for creating microstructures for a wide range of applications including, for example, photonic crystal structures.

Table 2. Three-dimensional optical data storage using photopolymerization

Author	Material	Objective	$\lambda(nm)$
Strickler, Webb [12]	Cibatool	60 × 1.4	620
Wang, Esener [15]	(see reference)	40 × 0.6	488
Cumpston et al. [16]	(see reference)	N/A	N/A
Sun et al. [17]	Nopocure 800	100 × 1.35	400
Maruo et al. [18]	SCR 500	60 × 0.85	790

2.4 Photobleaching Effect

Bhawalker et al. [19] reported on the abilities of high-efficiency two-photon excitation in a new fluorescent material. A large two-photon absorption cross-section fluorophore is required to generate efficient fluorescence with a given wavelength of light. In the case of two-photon photobleaching data storage, the fluorophore is doped into a polymer block. Illuminating the sample with an appropriate laser wavelength and average power (typically $< 1\,\mathrm{mW}$) from an ultra-short pulsed laser will produce two-photon fluorescence. Increasing the power above the bleaching threshold will cause the fluorophore to breakdown (bleach) and stop fluorescing.

By the use of this method a series of bleached patterns can be recorded in the material. The high localization of the fluorescence in the transverse and axial directions, known as optical sectioning, enables multiple layers to be recorded in the depth direction with a small layer spacing, resulting in a high capacity. The recorded information is read back using a two-photon fluorescence scanning microscope with the illumination power reduced to below the bleaching threshold.

Unfortunately, the information recorded using this method is permanent. Also, subsequent reading may photobleach the background, reducing the contrast of the recorded information, ultimately leading to an ineffective recording material.

Table 3 covers the different photobleaching polymers and equipment that have been used to record information in three dimensions.

Table 3. Three-dimensional optical data storage using photobleaching polymers

Author	Material	Objective	λ(nm)
Shih et al. [20]	APSS	40×1.3	800
Pan et al. [21]	APSS	40×1.3	800
Day et al. [22]	APSS	40×0.75	800
Pudavar et al. [23]	AF240	60×1.4	800

2.5 Photochromic Effect

Photochromism is the change of the molecular structure with a corresponding change in absorption upon illumination of an appropriate wavelength of light. The original lower energy state of the material is termed isomer 1, and the slightly higher energy state is referred to as isomer 2.

Parthenopoulos and *Rentzepis* [9,24] reported the three-dimensional recording of information in Spirobenzopyran (SP) using 'virtual' two-photon excitation. To achieve the energy required for 'virtual' two-photon excitation they used two orthogonal beams at wavelengths of 1064 nm and 532 nm, which when overlapped both spatially and temporally, excite at 400 nm. This differs slightly from the two-photon excitation process described in Sect. 2.2, where a single beam with a wavelength of half the energy required is focused into the sample. This second method for recording has been demonstrated in photochromic materials by *Toriumi* et al. [25].

The ability of photochromic compounds to transfer between isomer states makes it ideal for use in optical data storage. However, the ground state energy of the second isomer is typically slightly higher than the ground state energy of isomer 1. As a result of the difference in energy between the two ground states, there is a probability that molecules can thermally relax back to the ground state of isomer 1, thereby destroying the recorded information.

There exists a couple of methods by which the information can be read from photochromic materials. The first is detecting the fluorescence signal from isomer 2 [9], and the second is to detect the change in refractive index of the recorded bits [26]. This differs from the fluorescence reading system described in Sect. 2.4 for photobleaching materials, where the fluorescence from the background is read. A problem with this method of reading is that there is erasure of the recorded information, as it can convert back to isomer 1.

The second reading method involves detecting the change in refractive index of the recorded bit. Changing the chemical structure of the compound means that there is a slight change in the refractive index. The advantage of this method for reading is that a wavelength can be used to read the information that is absorbed by neither isomer 1 nor isomer 2. This method reduces the probability of information deterioration and improves the stability of the optical data bits.

Table 4 covers the different materials and equipment that have been used to record information within three dimensions using a photochromic polymer.

Table 4. Three-dimensional optical data storage using a photochromic polymer

Author	Material	Objective	λ(nm)
Parthenopoulos et al. [9]	SP	N/A	1064/532
Parthenopoulos et al. [24]	SP	N/A	1064/532
Toriumi et al. [27]	NSP	40×0.85	441.6
Toriumi et al. [25]	B1536	100×1.4	760

3 Photorefractive Effect

A photorefractive material has the ability to detect and store the spatial distribution of an optical intensity pattern as a change in the refractive index.

The photogenerated charges create a space-charge distribution, which produces an internal electric field that alters the refractive index by the Pockels effect (electro-optic effect) [28]. Traditionally crystals, such as $LiNbO_3$, were the only photorefractive materials. However, recently there has been a great deal of development in producing polymer based materials which exhibit the photorefractive effect.

3.1 Photorefractive Crystals

The mechanism outlined in this section is used to describe the photorefractive effect in crystals, although the internal behavior differs only slightly from that of the polymers.

The band transport model for the photorefractive crystal Fe:$LiNbO_3$ shows that upon absorption of a photon an electron is excited from the donor level to the conduction band. The free electrons then diffuse through the material, where they eventually recombine with an Fe^{3+} trap. As a result of the position-dependent space-charge distribution, an electric field is formed, which in turn modulates the refractive index.

When a photorefractive material is illuminated by an intensity distribution $I(r)$, which varies in the r direction, the absorption of a photon creates a free-charge distribution [28]. For the case of photorefractive crystals, an electron is excited from the donor level to the conduction band. The next step involves the diffusion and recombination of the free charges. Given that the illumination $I(r)$ is non-uniform, the number density of the free charges, $N(r)$, is also non-uniform. This produces regions of high concentrations of like charges. As a result the charges diffuse to areas of low concentrations. The non-uniform distribution of charges creates a position-dependent electric field $E(r)$. As the material is electro-optic, the internal electric field modifies the refractive index according to the Pockels effect, resulting in the following change in the refractive index $\Delta n(r)$ [28]:

$$\Delta n(r) = -\frac{1}{2} n^3 r_{\mathrm{e}} \frac{K_{\mathrm{b}} T}{e} \frac{1}{I(r)} \frac{\mathrm{d}I}{\mathrm{d}r} , \tag{2}$$

where n is the refractive index, r_{e} is the electro-optic coefficient of the material, K_{b} is the Boltzmann constant and T is the temperature.

Unlike photorefractive crystals, the photorefractive polymers do not yet have a defined charge transport mechanism. While the band transport model can work under certain conditions (i.e. no applied DC field), it cannot be used explicitly as a general model of the behavior of organic photorefractive polymer compounds [29,30].

For non-crystalline polymers the mobility of charges is severely limited as a result of the disordered structure of the compound compared with crystals. Consequently, the accuracy of the band transport model is reduced when the photorefractive effect in polymer-based materials is considered. A charge hopping model, where charges travel by hopping through side-chains or guest molecules appears to be the best model to describe a polymer-based photorefractive material [31].

There are four processes that should take place in order to produce a change in refractive index in an organic photorefractive polymer material: generation of charge carriers by illumination, transport of the charge carriers, trapping of the charge carriers in the dark regions and a linear electro-optic response.

The first physical process is the generation of mobile charges in response to the illumination profile. This may be viewed as the separation of electrons and holes [29] induced by the absorption of light in organic materials. A photosensitive compound is normally doped into the polymer host to provide the charge generation.

The second process is the transportation of one type of the charges, which in the case of organic materials is more likely to be the holes. If the mobility of the electrons and the holes was the same then the net effect would be to cancel the internal space-charge field created by this process. The two physical processes giving rise to charge transport are diffusion due to density gradients or drift due to an applied field. As most polymer materials with reasonable optical transparency are relatively good insulators, the ability of the charges to move by diffusion alone is limited. The dominant mechanism for charge transportation would be drift, which stimulates the charges to hop from one transport molecule to another. Typically the polymer is used as the charge transportation medium.

The third effect required for the photorefractive effect is trapping of the mobile charges. This effect is especially important when the lifetime of the recorded information is considered. A trapping site can be considered to be a region in the material where the charge is no longer able to move. The time that the charge is immobile is determined by the depth of the trap compared to the energy that the charge gains through absorption of incident illumination or from heating of the sample. The thermal energy of the charges at room temperature is one of the main causes of erasure of the recorded photorefractive pattern with time.

The final element necessary for the photorefractive effect is the linear electro-optic effect. The electro-optic effect modulates the refractive index as a result of the internal space-charge field. A nonlinear optical chromophore is normally doped into the polymer host to provide the electro-optic effect.

3.2 Localized Photorefractive Effect

The photorefractive effect is based on the non-uniform space-charge distribution produced, for example, by two intersecting beams as illustrated in Fig. 7. The sinusoidal interference pattern produced by the two beams is a result of the sum of the two intensity patterns [28]. This concept can be

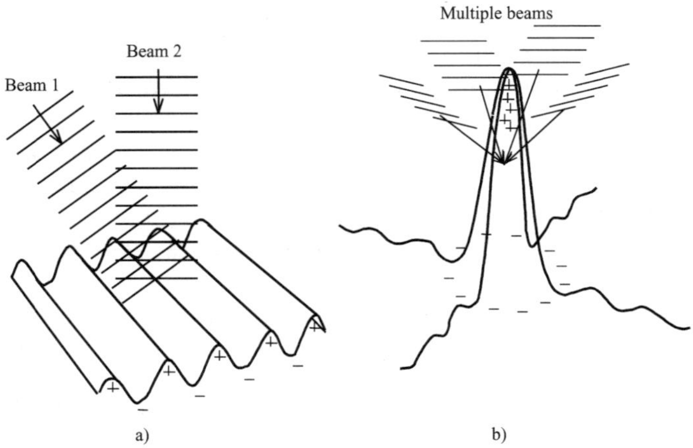

Fig. 7. (**a**) Interference pattern produced by two intersecting waves. (**b**) Diffraction pattern of an objective, which corresponds to the interference pattern produced from multiple beams intersecting in a circularly symmetric fashion

applied to the interference pattern produced from multiple beams. Increasing the number of intersecting beams in a circularly symmetric fashion creates an Airy pattern [28], corresponding to the diffraction pattern of an objective. Here the initial illumination from a focused beam produces a non-uniform charge distribution corresponding to the Airy pattern. After diffusion and recombination of the free charges (Fig. 8), a position-dependent internal space-charge field is produced. Finally, the presence of a nonlinear chromophore in a non-uniform space-charge field modulates the refractive index via Pockels effect [28].

3.3 Three-Dimensional Photorefractive Bit Data Storage

As described before, a photorefractive material is one that undergoes a change in refractive index as a result of a non-uniform illumination. As the change in refractive index is caused by an internal electric field, the information is easily erased by irradiating the sample with uniform illumination to produce a uniform distribution of the charges, and therefore remove any change in the refractive index associated with a recorded bit.

Previous work by *Kawata* et al. [32,33] has used single- and two-photon excitation to record information within the photorefractive $LiNbO_3$ crystal. *Hisaka* et al. [34] was also successful at recording in a Ce-doped SBN:75 crystal using domain reversal single-photon excitation.

As yet, photorefractive polymers have only been used in holographic data storage and two beam coupling experiments [35,36,37,38,39,40,41]. There are several advantages to using polymer-based materials for data storage. First, manufacturing of polymers into large recording samples is relatively easy and

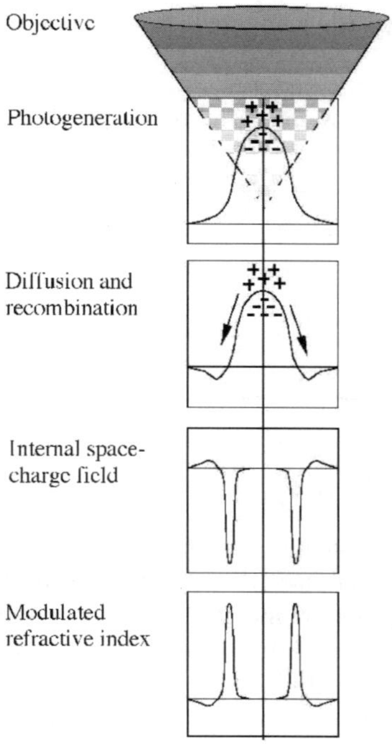

Fig. 8. Photorefractive mechanism for a focused beam

fast compared to growing highly doped crystals. Large-scale fabrication facilities already exist around the world to produce today's CDs and DVDs. It seems logical then that the next generation of optical data storage should be compatible with these technologies. Second, as a result of the difficulties in manufacturing high-quality inorganic crystals there exists a large difference in the production cost of a sample, not only making it expensive but also impractical as a commercial product. The final advantage is that there is room for improving the performance of the photorefractive polymers by tailoring the dopant compounds to suit the recording and reading systems. One such change would be to increase the absorption within a specific wavelength region used for recording, thereby reducing the laser power required to record information.

As a result, it is necessary to investigate three-dimensional bit optical data storage using photorefractive polymers.

Table 5 covers the different materials and equipment that have been used to record information within three dimensions using photorefractive materials.

Table 5. Three-dimensional optical data storage using inorganic photorefractive materials

Author	Material	Objective	λ(nm)
Ueki et al. [42]	Fe:$LiNbO_3$	40×1.0	476.5
Kawata et al. [32]	$LiNbO_3$	40×1.0	476.5
Kawata et al. [33]	$LiNbO_3$	40×0.85	760
Hisaka et al. [34]	Ce:SBN:75	40×0.75	488

4 Conclusions

It can be seen that the current trend in information storage requirements will continue to grow, while society gradually increases the quantity of information shared electronically. At present magnetic memories still offer vastly superior storage capacities; however, the transportability of optical memory provides society with a means of storing and distributing large quantities of information. The need for CDs and DVDs was driven by the music and movie industries, respectively, and as yet the only driver that appears to require an ever-increasing amount of information storage is the Internet.

All of the technologies that are being developed with the hope of taking optical storage systems into the future have their advantages and disadvantages. Data transfer speed is an important issue as the amount of information stored is increased; it is here that holographic storage has no equal. If miniaturization is the way of the future than solid immersion lens recording will provide a means for scaling down the storage mediums and maintaining the same amount of information. However, one key issue is the ability of a new technology to be compatible with the existing technology. Three-dimensional bit storage is capable of recording three orders of magnitude more information than is currently available on a DVD and yet will still be able to read a DVD.

References

1. Encyclopedia Britannica (2000)
2. T. Higuchi, S. Miyanabe, M. Okano: 27.4 Gbyte read-only dual-layer disc for blue laser, Proc. SPIE **3864**, 231–233 (1999)
3. K. Hirota, T. D. Milster, K. Shimura, Y. Zhang, J. S. Jo: Nearfield phase optical recording using a GaP hemispherical lens, Proc. SPIE **3864**, 361–363 (1999)
4. W. H. Yeh, M. Mansuripur: Evanescent coupling in magneto-optical and phase-change disk systems based on the solid immersion lens (SIL), Proc. SPIE 358–360 (1999)

5. T. D. Milster, K. Shimura, J. S. Jo, K. Hirota: Pupil-plane filtering for improved signal detection in an optical data storage system incorporating a solid immersion lens, Opt. Lett. **24**, 605–607 (1999)
6. L. D'Auria, J. P. Huignard, C. Slezak, E. Spitz: Experimental holographic read-write memory using 3-D storage, Appl. Opt. **13**, 808–818 (1974)
7. M. M. Wang, S. C. Esener, F. B. McCormick, I. Cokgoĉ, A. S. Dvornikov, P. M. Rentzepis: Experimental characterization of a two-photon memory, Opt. Lett. **22**, 558–560 (1997)
8. T. Tanaka, S. Kawata: Comparison of recording densities in three-dimensional optical storage systems: multilayered bit recording versus angularly multiplexed holographic recording, J. Opt. Soc. Am. A **13**, 935—943 (1996)
9. D. A. Parthenopoulos, P. M. Rentzepis: Three-dimensional optical storage memory, Science **245**, 843–845 (1989)
10. C. F. Bohern, D. R. Huffman: *Absorption and Scattering of Light by Small Particles* (Wiley, New York 1983)
11. X. S. Gan, M. Gu: Fluorescence microscopic imaging through tissue-like turbid media, J. Appl. Phys. **87**, 3214–3221 (2000)
12. J. H. Strickler, W. W. Webb: Three-dimensional optical data storage in refractive media by two-photon point excitation, Opt. Lett. **16**, 1780–1782 (1991)
13. W. Denk, J. H. Srickler, W. W. Webb: Two-photon laser scanning fluorescence microscopy, Science **248**, 129–131 (1990)
14. Y. R. Shen: *Principles of Nonlinear Optics* (Wiley, New York 1984)
15. M. M. Wang, S. C. Esener: Three-dimensional optical data storage in a fluorescent dye-doped photopolymer, Appl. Opt. **39**, 1826–1834 (2000)
16. B. H. Cumpston, S. P. Ananthavel, S. Barlow, D. L. Dyer, J. E. Ehrlich, L. L. Erskine, A. A. Heikal, S. M. Kuebler, I. Y. Sandy Lee, D. McCord-Maughon, J. Qin, H. Röckel, M. Rumi, X. L. Wu, S. R. Marder, J. W. Perry: Two-photon polymerization initiators for three-dimensional optical data storage and microfabrication, Nature **398**, 51–54 (1999)
17. H. B. Sun, S. Matsuo, H. Misawa: Three-dimensional photonic crystal structures achieved with two-photon-absorption photopolymerization of resin, Appl. Phys. Lett. **74**, 786–788 (1999)
18. S. Maruo, O. Nakamura, S. Kawata: Three-dimensional microfabrication with two-photon-absorbed photopolymerization, Opt. Lett. **22**, 132–134 (1997)
19. J. D. Bhawalker, J. Swiatkiewicz, S. J. Pan, J. K. Samarabandu, W. S. Liou, G. S. He, R. Berezney, P. C. Cheng, P. N. Prasad: Three-dimensional laser scanning two-photon fluorescence confocal microscopy of polymer materials using a new, efficient upconverting fluorophore, Scanning **18**, 562–566 (1996)
20. A. Shih, S. J. Pan, W. S. Liou, M. S. Park, J. D. Bhawalker, J. Swiatkiewicz, P. N. Prasad, P. C. Cheng: Three-dimensional image storage using two-photon induced photobleaching method, Cell Vision **4**, 223–224 (1997)
21. S. J. Pan, A. Shih, W. S. Liou, M. S. Park, J. D. Bhawalker, J. Swiatkiewicz, J. K. Samarabandu, P. Prasad, N. P. C. Cheng: Three-dimensional image recording by two-photon bleaching method, Scanning **19**, 156–158 (1997)
22. D. Day, M. Gu: Effect of refractive-index mismatch on three-dimensional optical data storage density in a two-photon bleaching polymer, Appl. Opt. **37**, 6299–6304 (1998)

23. H. E. Pudavar, M. P. Joshi, P. N. Prasad, B. A. Reinhardt: High-density three-dimensional optical data storage in a stacked compact disk format with two-photon writing and single photon reading, Appl. Phys. Lett. **74**, 1338–1340 (1999)
24. D. A. Parthenopoulos, P. M. Rentzepis: Two-photon volume information storage in doped polymer systems, J. Appl. Phys. **68**, 5814–5818 (1990)
25. A. Toriumi, S. Kawata, M. Gu: Reflection confocal microscope readout system for three-dimensional photochromic optical data storage, Opt. Lett. **23**, 1924–1926 (1998)
26. Y. Kawata, R. Jukaitis, T. Tanaka, T. Wilson, S. Kawata: Differential phase-contrast microscope with split detector for the readout system of a multilayered optical memory, Appl. Opt. **35**, 2466–2470 (1996)
27. A. Toriumi, J. M. Herrmann, S. Kawata: Nondestructive readout of a three-dimensional photochromic optical memory with a near-infrared differential phase-contrast microscope, Opt. Lett. **22**, 555–557 (1997)
28. B. E. A. Saleh, M. C. Teich: *Fundamentals of Photonics* (Wiley, New York 1991)
29. W. E. Moerner, S. M. Silence: Polymeric photorefractive materials, Chem. Rev. **94**, 127–155 (1994)
30. W. E. Moerner, A. Grunnet-Jepsen, C. L. Thompson: Photorefractive polymers, Annu. Rev. Mater. **27**, 585–623 (1997)
31. C. Bosshard, K. Sutter, P. Prêtre, J. Hulliger, M. Flörsheimer, P. Kaatz, P. Günter: *Organic Nonlinear Optical Materials* (Gordon and Breach, Switzerland 1995)
32. Y. Kawata, H. Ueki, Y. Hashimoto, S. Kawata: Three-dimensional optical memory with a photorefractive crystal, Appl. Opt. **34**, 4105–4110 (1995)
33. Y. Kawata, H. Ishitobi, S. Kawata: Use of two-photon absorption in a photorefractive crystal for three-dimensional optical memory, Opt. Lett. **23**, 756–758 (1998)
34. M. Hisaka, H. Ishitobi, S. Kawata: Optical recording of reversed domains in a Ce-doped SBN:75 crystal for bit oriented three-dimensional optical memory, J. Opt. Soc. Am. B **17**, 422–426 (2000)
35. K. Meerholz, B. L. Volodin, D. Sandalphon, B. Kippelen, N. Peyghambarian: A photorefractive polymer with a high optical gain and diffraction efficiency near 100%, Nature **371**, 497–500 (1994)
36. R. Birabassov, N. Landraud, T. V. Galstyan, A. Ritcey, C. G. Bazuin, T. Rahem: Thick dye-doped poly(methyl methacrylate) films for real-time holography, Appl. Opt. **37**, 8264–8269 (1998)
37. R. Bittner, Brèuchle, C. K. Meerholz: Influence of the glass-transition temperature and the chromophore content on the grating buildup dynamics of poly(N-vinylcarbazole)-based photorefractive polymers, Appl. Opt. **37**, 2843–2851 (1998)
38. J. J. Stankus, S. M. Silence, W. E. Moerner, G. C. Bjorklund: Electric-field-switchable stratified volume holograms in photorefractive polymers, Opt. Lett. **19**, 1480–1482 (1994)
39. K. Matsushita, P. P. Banerjee, S. Ozaki, D. Miyazaki: Multiwave coupling in a high-gain photorefractive polymer, Opt. Lett. **24**, 593–595 (1999)
40. V. P. Pham, G. Manivannan, R. A. Lessard: Holographic characterization of azo-dye-doped poly(methyl methacrylate) films, Thin Solid Films **270**, 295–299 (1995)

41. V. P. Pham, G. Manivannan, R. A. Lessard, G. Bornengo, R. Po: New azo-dye-doped polymer systems as dynamic holographic recording media, Appl. Phys. A **60**, 239–242 (1995)
42. H. Ueki, Y. Kawata, S. Kawata: Three-dimensional optical bit memory recording and reading with a photorefractive crystal: analysis and experiment, Appl. Opt. **35**, 2457–2465 (1996)

Two-Step Processes and IR Recording in Photorefractive Crystals

Eckhard Krätzig[1] and Karsten Buse[2]

[1] Department of Physics, University of Osnabrück
Barbarastr. 7, 49069 Osnabrück, Germany
eckhard.kraetzig@physik.uni-osnabrueck.de

[2] Institute of Physics, University of Bonn
Wegelerstr. 8, 53115 Bonn, Germany
kbuse@uni-bonn.de

Abstract. Two-step excitation processes have been used for hologram storage in photorefractive crystals. By this means the interference pattern can be formed with red or near–IR light and nondestructive readout of information is possible. Often shallow levels are involved in the holographic recording process in photorefractive crystals. The shallow levels can be populated by illumination with visible or UV pulses forming states with relatively long lifetimes, thus sensitizing the crystals for holographic recording with IR pulses. In $LiNbO_3$ and $LiTaO_3$ the most important shallow levels have been identified. They result from $Nb_Li{}^{5+}$ and $Ta_Li{}^{5+}$ antisite defects (Nb^{5+} or Ta^{5+} on Li^+ site). The crystals can also be pre-illuminated with visible light from a cw argon laser or a xenon lamp and holograms can be recorded with red light from a laser diode. The sensitization process is possible for other photorefractive crystals, too. The holograms can be read nondestructively with IR light and can be erased with green light. The hologram lifetime is limited by electron tunneling or by an ionic conductivity. Lifetimes up to years can be achieved. Recording of components for telecommunication applications with IR light allows one to create reconfigurable and thus more versatile devices.

1 Introduction

Storage of volume phase holograms in electro-optic crystals like $LiNbO_3$ offers fascinating possibilities for many applications [1,2]. The involved photorefractive effect is based on the transposition of a light pattern into a refractive index pattern. Under nonuniform illumination charge carriers – electrons or holes – are excited and trapped at new sites. By this means electrical space-charge fields are built up which give rise to an electro-optic modulation of the refractive index. The trapped charge can be released and the field pattern erased by uniform illumination or by heating.

But there are two main drawbacks. First, the crystals are insensitive in the interesting near-IR region. Second, retrieval of stored information requires homogeneous illumination and thus necessarily leads to erasure effects.

To overcome these drawbacks the use of two-photon excitation has been proposed for hologram recording by *von der Linde* et al. [3]. Then the holograms can be recorded with near-IR pulses. Furthermore, readout without

P. Boffi, P. Piccinin, M. C. Ubaldi (Eds.): Infrared Holography for Optical Communications, Topics Appl. Phys. **86**, 23–39 (2003)

erasure is possible using pulses of reduced light intensity. The Bragg condition is fulfilled because the light wavelength remains unchanged, but the energy of one photon is not sufficient to excite an electron to the conduction band and thus redistribution of electrons does not occur. On the other hand, optical erasure may be performed with the help of two-photon processes.

A serious disadvantage of two-photon recording using virtual intermediate states is the necessity of extremely high peak intensities of the light pulses. If impurity ions exhibiting real intermediate states with long lifetimes can be utilized, a considerable reduction of the peak intensity required for two-photon recording is possible, though the intensity has to be still much larger than in the case of usual one-step recording.

In the 1990s it became clear that shallow levels are often involved in the holographic recording process in photorefractive crystals. The shallow levels can be populated by illumination with visible or UV light, forming states with relatively long lifetimes, thus sensitizing the crystals for holographic recording with IR light.

In the present contribution two-step processes for holographic recording in different photorefractive crystals are investigated. We discuss the use of near-IR light and nondestructive readout. Models for the underlying physical processes are presented. Recent data about hologram lifetimes will be shown and general considerations for the application of two-step methods with the fabrication of telecommunication components will be discussed.

2 Early Experiments

First two-step excitation measurements in photorefractive crystals were performed in 1974 [3]. Frequency-doubled light pulses of a mode-locked Nd:YAG laser (wavelength 0.53 μm, pulse duration 10 ps) induced a phase retardation in doped $LiNbO_3$ crystals. It was found that the measured refractive-index changes depend quadratically on the exposure energy. The influence of additional IR pulses (1.06 μm) was demonstrated, too. Furthermore, $LiNbO_3$:Cr and $LiTaO_3$:Cr crystals have been investigated [4] to utilize the long-lived $^4T_2$ excited state of Cr^{3+} (lifetime 500 ns). In this case the peak intensity is reduced and pulse lasers with higher repetition rates can be used.

We have demonstrated nondestructive readout in $LiTaO_3$:Fe [5]. Holograms were recorded by simultaneous illumination with 30 ps pulses of a mode-locked Nd:YAG laser at 1.06 μm forming the interference pattern and with spatially homogeneous frequency-doubled pulses of the same laser at 0.53 μm. Results are shown in Fig. 1. The holograms can be read nondestructively at 1.06 μm. From the experimental accuracy it was concluded that more than 10000 readout processes are possible. Similar results were obtained for $LiNbO_3$:Cr using 40ns pulses of a Q-switched ruby laser at 0.694 μm [6].

In the 1990s it became clear that – at least at high light intensities – shallow levels are nearly always involved in the holographic recording process

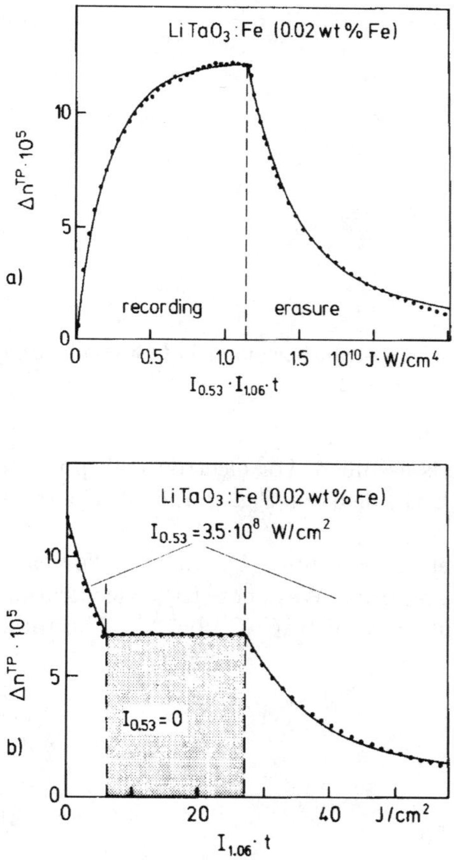

Fig. 1. (**a**) Two-photon recording-erasure cycle. The refractive-index amplitude $\Delta n^T P$ is plotted versus the product $I_{0.53} \times I_{1.06} \times t$ of green and IR intensities and time. (**b**) Experiment as in (**a**), but now $\Delta n^T P$ is plotted only versus the product $I_{1.06} \times t$ of IR intensity and time. Without green light ($I_{0.53} = 0$, hatched region) no erasure is observed

in photorefractive crystals. Two-level models – the two-center and the three-valence model – were developed and successfully applied for the description of the light-induced charge transport in many photorefractive crystals.

A very important step for the understanding of the photorefractive properties of ferroelectric perovskites was the discovery of light-induced absorption in $BaTiO_3$ by *Motes* and *Kim* [7]. This increase of absorption under illumination was interpreted in terms of two kinds of centers involved [8], each of them occurring in two different states. *Holtmann* successfully applied this two-center model to describe the transport properties of $BaTiO_3$ [9].

Because the photoconductivity of ferroelectric perovskites is mostly dominated by holes in the valence band [9,10] in the following discussion of the

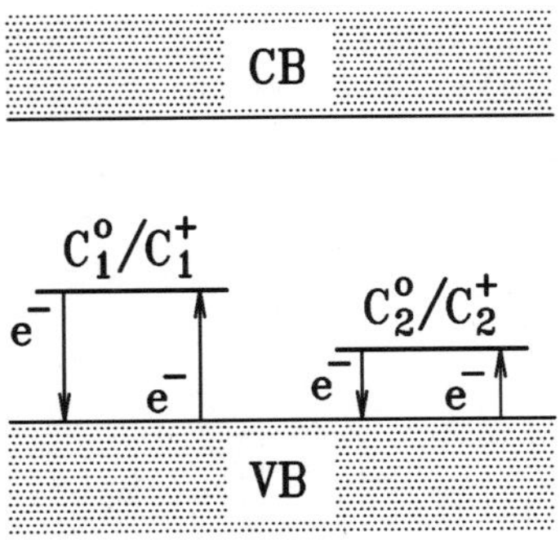

Fig. 2. Band diagram of the two-center charge-transport model (CB: conduction band, VB: valence band, C_1: center 1, C_2: center 2)

two-center model only hole transport is assumed. For electron transport an analogous argumentation holds. With the help of Fig. 2 the transport of charge may be described as follows.

We consider two different photorefractive centers C_1 and C_2. For each species, $i = 1, 2$, there are hole sources and traps. We denote the concentration of sources C_i^+ by N_i^+ and the concentration of traps C_i^0 by N_i^0. The total concentration of centers of type i is

$$N_i = N_i^+ + N_i^0 \quad . \tag{1}$$

Charge conservation requires

$$N_1^+ + N_2^+ + N_h = N_c \quad , \tag{2}$$

where N_h is the concentration of holes in the valence band and N_c a constant concentration. The first center has to be a deep-level impurity, e.g. iron, and the second one should be a more shallow trap with respect to the valence-band edge. The center C_2 has a relatively low thermal activation energy, such that $N_2^0 \gg N_2^+$ holds in the dark. Upon illumination holes are generated by excitation of electrons from the valence band into C_1^+ centers. The holes migrate in the valence band and are trapped either by C_1^0 or by C_2^0 centers. Trapping at the latter creates C_2^+ centers. With increasing light intensity, more and more holes are generated and N_2^+ grows, too. By this means absorption processes become possible which result from optical excitations of valence band electrons to C_2^+ centers. This leads to light-induced absorption changes, if C_1^+ and C_2^+ have different photon-absorption cross-sections. The rate equations read:

$$\frac{dN_i^+}{dt} = -(q_i S_i I + \beta_i) N_i^+ + r_i (N_i - N_i^+) N_h, \quad i = 1, 2 \quad . \tag{3}$$

Here q_i denote the quantum efficiencies for generating a hole upon absorption of a photon, S_i the absorption cross-sections, β_i the thermal generation rates and r_i the recombination coefficients. Many experimental results can be understood on the basis of this two-center model, among them the nonlinear dependence of the photoconductivity on the light intensity; details are described in [11].

But there exists another model for explaining the charge transport properties of perovskites. As we have pointed out [12], the assumption of one impurity center occurring in three different valence states – the so-called three-valence model – leads to conclusions that are similar to those of the two-center model.

The situation is illustrated in Fig. 3.

The three valence states of the center C are denoted by 0, + and 2+. The arrows indicate the considered excitation and recombination processes of electrons. At low intensities only C^0 and C^+ states are present, because thermally excited valence band electrons fill C^{2+}. Illumination excites electrons from the valence band into C^+ and generates holes which are annihilated by electrons from C^0. For sufficiently high light intensities the hole concentration becomes large enough that an appreciable number of electrons from C^+ can recombine with holes and generate C^{2+} contributing to absorption. Thus light-induced absorption changes appear. Furthermore, participation of C^{2+} in the charge transport may provide a photoconductivity increasing nonlinearly with light intensity.

Rate equations, charge conservation and constant trap density may be written as:

$$\frac{dN^+}{dt} = +r^0 N^0 N_h - (\beta^+ + q^+ S^+ I) N^+ \\ -r^+ N^+ N_h + (\beta^{2+} + q^{2+} S^{2+} I) N^{2+}\,, \tag{4}$$

$$\frac{dN^{2+}}{dt} = +r^+ N^+ N_h - (\beta^{2+} + q^{2+} S^{2+} I) N^{2+}\,, \tag{5}$$

$$2\,\mathrm{N}^{2+} + N^+ + N_h = N_c\,, \tag{6}$$

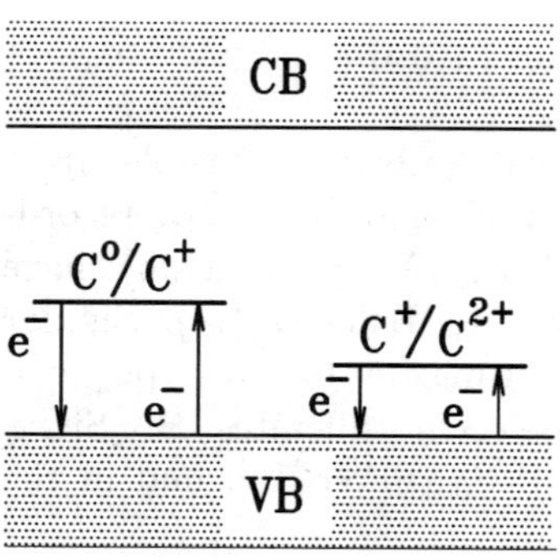

Fig. 3. Band diagram of the three-valence model (CB: conduction band, VB: valence band; the valence states of the center C are indicated by 0, + and 2+)

$$N^0 + N^+ + N^{2+} = N \,. \tag{7}$$

Here again N^0, N^+ and N^{2+} are the concentrations of C^0, C^+ and C^{2+}, N is the entire impurity concentration, N_h is the concentration of holes in the valence band, β^+ and β^{2+} are the thermal generation rates, q^+ and q^{2+} are the quantum efficiencies for hole generation upon absorption of a photon, S^+ and S^{2+} are the photon absorption cross-sections, r^0 and r^+ are the recombination coefficients and N_c is a constant concentration. It should be emphasized that the above equations of the three-valence model cannot be derived from those of the two-center model introducing a special relation between the concentrations of deep and shallow traps. But both models lead to similar conclusions.

Though it is often difficult to decide between the two-center and the three-valence model [12], in both cases shallow levels are involved which may be utilized for two-step excitations.

These models that were developed originally for perovskite-type crystals can be applied to other materials as well. One special case of the two-center model is the situation of two deep centers, i.e. the thermal generation rates β_i are both negligible. Such a system is, for example, $LiNbO_3$ doped with Mn and Fe, where $Mn^{2+/3+}$ is a deeper impurity than $Fe^{2+/3+}$. The material can be used for persistent optical data storage [13]. Homogeneous UV light excites electrons from Mn^{3+} into the conduction band. These electrons are partially trapped by Fe^{3+} forming Fe^{2+} centers. From these Fe^{2+} centers electrons can be excited by red or by green light. By this means the material is sensitized for red or green recording. The hologram is partially stored in the Mn and in the Fe centers. Reconstruction of the hologram with the reading light erases only the hologram in the Fe centers since the photon energy is not large enough to excite electrons from Mn^{2+}. Thus part of the hologram is resistant against reading. This method works with continuous-wave light since there are no competing thermal processes. But it is not useful for near-IR recording since two deep centers are involved.

3 Two-Step Excitation via Shallow Levels

Population of shallow levels with nanosecond pulses of visible light (frequency-doubled Nd:YAG laser, $\lambda = 0.53\,\mu m$) and hologram recording with nanosecond IR pulses (Nd:YAG laser, $\lambda = 1.06\,\mu m$) have been demonstrated in the ferroelectric perovskite $BaTiO_3$ [14]. Similar experiments have also been performed in the tungsten-bronze-type crystal $Sr_{1-x}Ba_xNb_2O_6$:Ce [15], in the nonferroelectric sillenites $Bi_{12}SiO_{20}$ (BSO) [16] and $Bi_{12}TiO_{20}$ (BTO) [17], in $Sn_2P_2S_6$ (SPS) [18], and in the semiconductors CdZnTe:V [19] and CdTe:Ge [20].

But the early experiments in $LiNbO_3$ and $LiTaO_3$ [3,5,6,21] also have to be discussed with respect to the new knowledge about the light-induced charge transport. In these materials shallow levels have been identified [22,23]. They

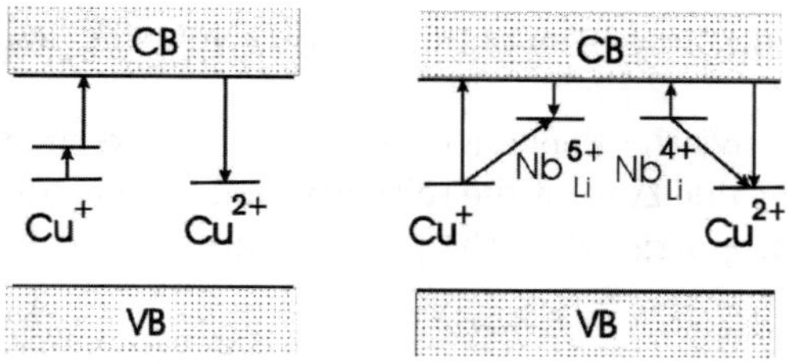

Fig. 4. Two possible two-step excitations in $LiNbO_3$:Cu, via excited states of Cu^+ (*left*) and via Nb_{Li}^{5+} shallow traps (*right*)

result from Nb_{Li}^{5+} and Ta_{Li}^{5+} antisite defects (Nb^{5+} or Ta^{5+} on Li^+ site), which are present in great quantities in congruent crystals. The highly charged Nb^{5+} or Ta^{5+} ions trap mobile electrons and form small polarons.

There are two possibilities for two-step excitations in these crystals: via excited states of the deep traps and via shallow traps. These possibilities are explained in Fig. 4.

To decide between the two possibilities, measurements with time-delayed pulses have been performed [24,25]. The results are summarized in Fig. 5 for $LiNbO_3$:Cu. The crystal is illuminated with a homogeneous green (0.53 µm) pulse and two intersecting IR (1.06 µm) pulses of about 20 ns.

In order to sensitize the crystal for IR holographic recording, previous or simultaneous illumination with the green light is necessary. If the green pulse impinges after the IR pulses, no refractive-index change is obtained. These results clearly indicate that indeed shallow traps are involved in the two-step excitation process; the photon energy of the IR light is not sufficient to pop-

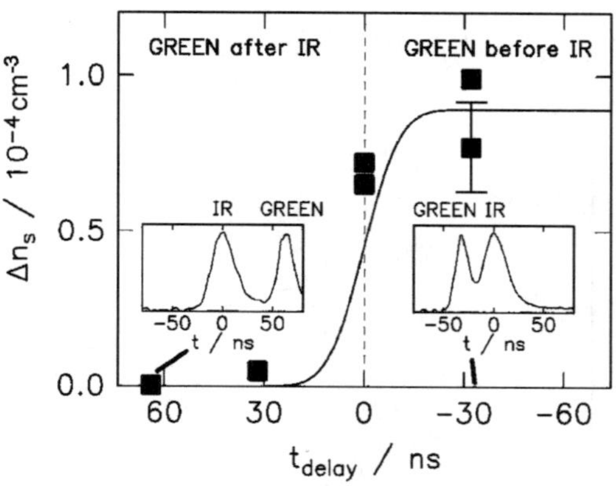

Fig. 5. Saturation value of refractive-index change Δn_s versus delay lime t_{delay} between IR and green pulses in $LiNbO_3$:Cu. The squares represent measured values and the solid line guides the eye. The insets illustrate the pulse intensity development for two different delay times (64 ns and −32 ns)

ulate shallow traps. Analogous results have been obtained for $LiNbO_3$:Fe [25] and $LiTaO_3$:Fe [26].

Furthermore, we have investigated the dependences of the inverse time constant τ^{-1} and of the saturation value Δn_s of the refractive index changes. The inverse time constant τ^{-1} is proportional to the photoconductivity σ_{ph},

$$\tau^{-1} = \sigma_{ph}/(\epsilon\epsilon_0) \quad , \tag{8}$$

where ϵ is the static dielectric constant and ϵ_0 the vacuum permittivity. The quantities τ^{-1} and Δn_s depend on the concentrations of the impurity ions involved and on the intensities of the IR and the visible light. Experimental results are shown in the Figs. 6 and 7.

From these results the following relations have been deduced, which are valid for $LiNbO_3$:Cu [24], $LiNbO_3$:Fe [25] and $LiTaO_3$:Fe [26]:

$$\sigma_{ph} \propto (c_D^f/c_D^e) I^{VIS} \quad , \tag{9}$$

$$\Delta n_s \propto c_D^e I^{IR} \quad , \tag{10}$$

where c_D^f denotes the the concentration of filled deep traps (Fe^{2+}, Cu^+), c_D^e the concentration of empty deep traps (Fe^{3+}, Cu^{2+}), I^{VIS} the intensity of

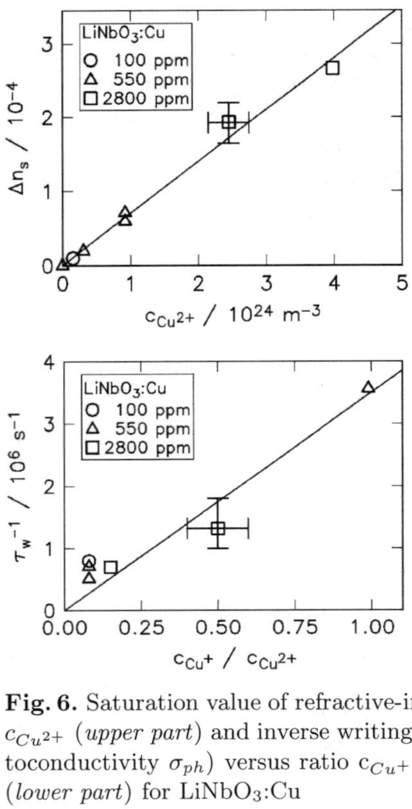

Fig. 6. Saturation value of refractive-index change Δn_s versus Cu^{2+} concentration $c_{Cu^{2+}}$ (*upper part*) and inverse writing time constant τ_w^{-1} (proportional to the photoconductivity σ_{ph}) versus ratio $c_{Cu^+}/c_{Cu^{2+}}$ of the Cu^+ and Cu^{2+} concentration (*lower part*) for $LiNbO_3$:Cu

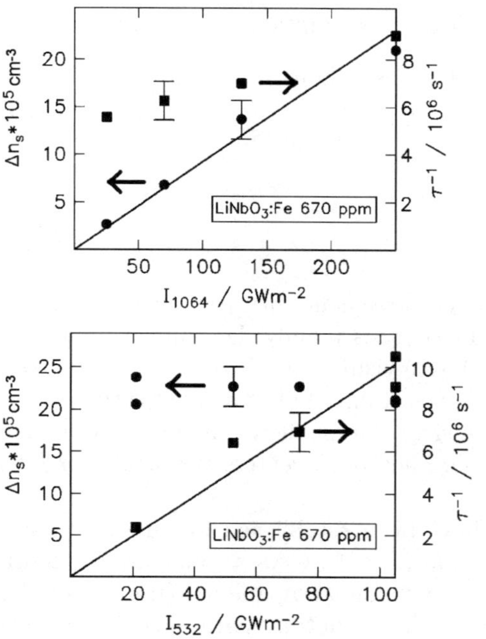

Fig. 7. Saturation value of refractive-index change Δn_s and inverse writing time constant τ^{-1} versus intensity I_{1064} of the IR light (upper part) and versus intensity I_{532} of the green light (lower part), respectively, for $LiNbO_3$:Fe

the visible light, I^{IR} the intensity of the IR light and Δn_s the saturation value of the refractive-index amplitude.

These relations can be derived from the above model assuming the participation of shallow levels [26] (Fig. 4, right-hand side). The light-induced charge transport in doped $LiNbO_3$ and $LiTaO_3$ is mainly determined by a bulk photovoltaic effect [27]. The photovoltaic current density j_{pv} contains the modulated intensity I^{IR} of the IR light:

$$j_{pv} = \kappa_{Sh}^{IR} I^{IR} c_{Sh}^{f} \quad , \tag{11}$$

where κ_{Sh}^{IR} is the photovoltaic constant and c_{Sh}^{f} the concentration of the filled shallow traps ($Nb_{Li}^{4+}, Ta_{Li}^{4+}$). The photoconductivity $\sigma_{ph} = e\mu_e c_e = e\mu_e g_e \tau_e$ (μ_e mobility, c_e concentration, g_e generation rate, τ_e lifetime of excited carriers) may result from excitations from filled deep traps (Fe^{2+}, Cu^{+}) and filled shallow traps (Nb_{Li}^{4+}, Ta_{Li}^{4+}):

$$\begin{aligned}\sigma_{ph} = e\mu_e (&S_D^{VIS} q_D^{VIS} I^{VIS} c_D^{f} + S_{Sh}^{VIS} q_{Sh}^{VIS} I^{VIS} c_{Sh}^{f} + S_{Sh}^{IR} q_{Sh}^{IR} I^{IR} c_{Sh}^{f}) \\ &\times (r_D c_D^{e} + r_{Sh} c_{Sh}^{e})^{-1} \quad ,\end{aligned} \tag{12}$$

where S denotes the cross-sections (for filled deep and shallow traps), q the quantum efficiencies (for filled deep and shallow traps) and r the recombination coefficients (for empty deep and shallow traps). The description of the experimental results requires the following assumptions:

$$\begin{aligned} \sigma_{ph} &= e\mu_e (S_D^{VIS} q_D^{VIS} I^{VIS} c_D^f)(r_D c_D^e)^{-1} \propto (c_D^f / c_D^e) I^{VIS} \quad , \\ j_{pv} &= \kappa_{Sh}^{IR} I^{IR} c_{Sh}^f \propto I^{IR} I^{VIS} c_D^f \quad . \end{aligned} \tag{13}$$

Therefore, we obtain

$$\Delta n_s \propto j_{pv} / \sigma_{ph} \propto c_D^e I^{IR} \quad . \tag{14}$$

Equations (13) and (14) are in perfect agreement with the experimental results, (9) and (10). The photoconductivity is mainly determined by excitations of filled deep traps (Fe^{2+}, Cu^{+}) with visible (or UV) light (homogeneous intensity), the photovoltaic effect by excitations of filled shallow traps (Nb_{Li}^{4+}, Ta_{Li}^{4+}) with IR light (modulated intensity). The shallow centers are populated by direct excitation of electrons from filled deep centers by visible (or UV) light ($c_{Sh}^f \propto I^{VIS} c_D^f$).

Excitations via shallow levels have been established for many crystals. Some samples may also be $LiNbO_3$ and $LiTaO_3$. Crystals may also be sensitized with visible light of a cw laser or a xenon lamp, and hologram recording is possible with a laser diode [28,29,30]. Two-photon excitations via intermediate states cannot be excluded in some cases, but have not yet been demonstrated unambiguously.

Holographic experiments have also been performed with laser pulses in the 100-fs range [31,32]. Undoped $BaTiO_3$ crystals are sensitive to these ultrashort pulses even in the 1.5- μm wavelength regime [31], but it is again difficult to decide whether excitations via intermediate states or via shallow traps are involved.

4 Lifetime of the Holograms

For practical applications in, for example, telecommunication the lifetime of the components is a critical issue. Wavelength filters that are based on holographic gratings in photorefractive $LiNbO_3$ crystals [33,34,35,36] utilize so far the method of 'thermal fixing' [37]. Holograms were recorded at a higher temperature (typically 180 °C), where ions are mobile. These ions migrate and compensate for the electronic space-charge field. The spatially modulated concentrations of filled and empty electron traps as well as a spatially modulated concentration of the compensating ions build up. After cooling to room temperature the ions are practically immobile. Homogeneous illumination now generates spatially modulated currents because of the modulated densities of filled and empty electron traps. Space-charge fields and electrooptic refractive index changes arise (see [38] and references therein). After reaching a steady-state, further net charge redistribution is not possible since the ionic grating is fixed.

From accelerated ageing experiments assisted by a theoretical description of the processes it can be shown that the lifetime of thermally fixed holograms in $LiNbO_3$ can reach hundreds of years at room temperature [39]. Usually protons (H^+) form the ionic grating [38,40]. The key to getting good lifetimes is a dehydration of the $LiNbO_3$ crystals [39,41]. Then the compensating ions are – most probably – lithium ions instead of hydrogen [41]. Since the mobility of Li^+ in $LiNbO_3$ is less than that of H^+, the lifetime of the thermally fixed gratings is improved. By this means the telecommunication lifetime standards (Bellcore, Telecordia) can be easily fulfilled.

However, thermal fixing has two drawbacks: (1) Optical erasure and rewriting of components is not possible. For many applications an all-optical control of the diffractive components is desirable. Just to name one example: guiding one light beam into one of many fibers can be done by recording of a proper hologram that provides this coupling. For switching the channel into another fiber a new hologram is required. Optical erasure of the old hologram and recording of a new hologram can provide such a reconfigurable switch. Thermally fixed holograms do not provide this flexibility. (2) Photorefractive crystals show in general an anisotropic thermal expansion. Thus, after recording and cooling to room temperature, typically the Bragg condition is not fulfilled anymore. This can be compensated by, for example, a change of the angle of incidence or of the wavelength of the reading light. However, this is limited to gratings only. For other waves Bragg matching cannot be achieved by a simple modification of the reading light. Waves with so-called 'wavevector spectra' might be one solution [42], but this limits the quality of the component, i.e. the focal spot of nominally spherical waves would be enlarged and the cross-talk for wavelength-division-multiplexing components would be increased. Another solution to this problem is the 'low-high-low' fixing schedule, where the hologram is recorded at room temperature and heated afterwards. However, for this method the obtainable refractive-index changes, i.e. the diffraction efficiencies, are smaller. To summarize this point: thermal fixing works well for applications where elementary gratings are involved but fails for more sophisticated components.

By two-step excitation processes these problems can be overcome. Such components are optically erasable and rewritable; the recording light always fulfills the Bragg condition. However, there is one drawback: The lifetime is not as good as that of thermally fixed holograms. Recent studies reveal the mechanisms that are responsible for the erasure of unfixed holograms in $LiNbO_3$ [43,44]. Because of the two-step recording process, reading with IR light does not erase the holograms. However, there is always a dark conductivity σ_d present that limits the hologram lifetime given by

$$\tau_{life} = (\epsilon\epsilon_0)/\sigma_d \quad . \tag{15}$$

The lifetime in the dark is the same for holograms recorded by one-step or by two-step processes.

Two situations need to be distinguished: for $LiNbO_3$ crystals containing a high amount of iron or copper (typically in excess of 0.05 wt.% Fe_2O_3 or CuO) the dark conductivity is dominated by tunneling of electrons between these deep impurities [43]. Figure 8a shows the dark conductivity of a $LiNbO_3$:Fe crystal where the Fe^{2+}/Fe^{3+} concentration ratio was changed by thermal annealing. As it can be seen, the dark conductivity is proportional to the effective trap density N_{eff} that is defined as $N_{\text{eff}} = (1/c_{\text{Fe}^{2+}} + 1/c_{\text{Fe}^{3+}})^{-1}$. For small Fe^{2+} concentrations $c_{\text{Fe}^{2+}}$, the number of electrons that can tunnel is given by $c_{\text{Fe}^{2+}}$ itself, while for high Fe^{2+} concentrations a lack of empty Fe sites, i.e. of Fe^{3+}, limits the tunneling, and hence the dark conductivity is proportional to the density $c_{\text{Fe}^{3+}}$. This explains the dependence $\sigma_d \propto N_{\text{eff}}$ and at the same time shows that indeed electron tunneling is responsible for the dark conductivity. Since the tunneling probability depends exponentially on the distance between the ions, the overall iron concentration also plays an important role. As Fig. 8b shows, the dark conductivity (normalized to N_{eff}) increases indeed exponentially with $(c_{\text{Fe}^{2+}} + c_{\text{Fe}^{3+}})^{1/3} = c_{\text{Fe}}^{1/3}$, the averaged reciprocal spacing between the iron centers. Based on this curve – for the case of iron-dominated conductivity – the dark storage time of the holograms can be deduced.

The second situation is that of a relatively low doping level. In this case the ionic conductivity at room temperature dominates the charge transport [43,44]. This conductivity can be reduced by dehydration of the crystal. Thus the best lifetimes are obtained in weakly doped crystals with a small H^+ content. However, weak doping implies small refractive-index changes, as can be inferred, for example from Fig. 6. For realistic conditions and after optimization, hologram lifetimes at room temperature of the order of months to years can be achieved. At higher temperatures, e.g. at 50 °C, lifetimes of only days or weeks are expected. This is not satisfactory for telecommunication applications. The consequence is that such components need either to be temperature stabilized or they must contain an apparatus that can rewrite or refresh the components on demand. For dynamic components this is not necessarily a drawback since they may contain in any case a recording laser.

5 Advantages of Infrared Recording

For single-step volatile recording in photorefractive crystals usually ultraviolet or visible light is required. Only in a few cases is single-step recording with near-IR light possible [45]: crystals of the sillenite type are sensitive to near-IR light, and two-step recording processes have been demonstrated in this material as well [17], but the dark storage time is not satisfactory for most applications. Semiconductor crystals like CdTe and GaAs are sensitive in the telecommunication wavelength region (1.3 to 1.6 μm), but the refractive-index changes are too small since the electro-optic coefficients are tiny. Only photorefractive multiple quantum wells provide large refractive-index changes for

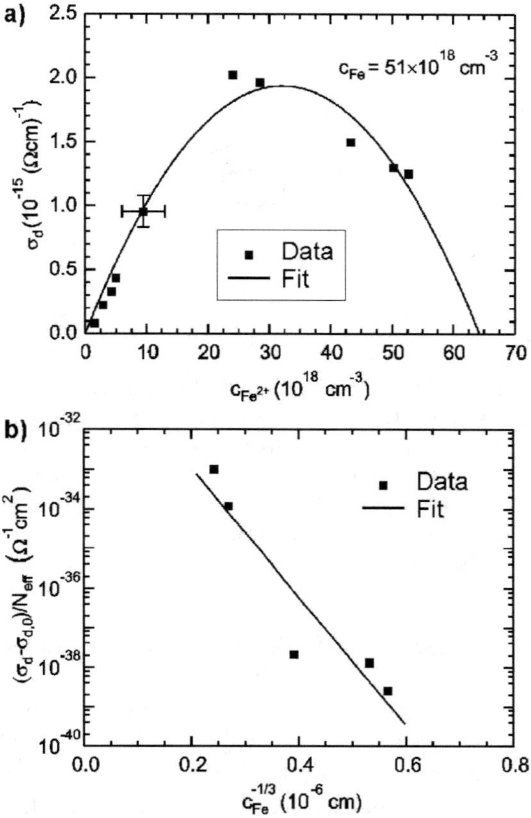

Fig. 8. (**a**) Dark conductivity σ_d versus Fe^{2+} concentration $c_{Fe^{2+}}$ of a $LiNbO_3$ sample doped with 51×10^{18} cm^{-3} Fe. The solid line shows a fit of $\sigma_d \propto N_{eff}$ to the experimental data, where N_{eff} is the effective trap density (see text). (**b**) Normalized dark conductivity σ_d versus the cubic root of the iron concentration c_{Fe}. Here $\sigma_{d,0}$ is an iron-independent background dark conductivity and N_{eff} is again the effective trap density. The *solid line* is a fit of conductivity $(\sigma_d - \sigma_{d,0})/N_{eff}$ versus $\exp(-c_{Fe}^{-1/3})$

recording in the telecommunication wavelength region, but the operational wavelength range is small. In very few cases, e.g. for highly doped $KNbO_3$ waveguides, a photorefractive response in the telecommunication region has been reported for other materials [46]. Thus two-step recording has – besides the resistance against reading with IR light – one still-more substantial advantage: devices can be fabricated with light of the wavelength where they are finally used. This is different compared to the current approach where gratings are recorded by, for example, green light and finally read by IR light [33,34,35,36]. Since volume effects are used, this approach is limited to

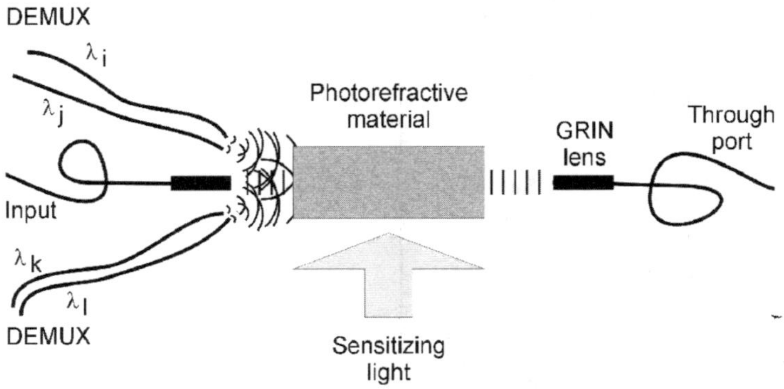

Fig. 9. Filter for demultiplexing (DEMUX) of four channels (λ_i, λ_j, λ_k, λ_l) in a wavelength-division-multiplexing network. Only the input and the through port require gradient-index (GRIN) lenses to collimate the beam. Holograms can be multiplexed in such a way that the diffracted light is coupled into the four separate fibers without any additional optics. The required holograms can be recorded in the device itself by a two-step process if sensitizing light is present. The sensitizing light is not required for operation of the device

gratings only. For other elements Bragg matching in the IR spectral range would be impossible.

Figure 9 shows that recording with IR light that has exactly the wavelength of the light used in the final device can be very helpful. A four-channel wavelength-demultiplexer (DEMUX) is depicted. In the input fiber many wavelengths, e.g. 128 or 256, might be multiplexed. A gradient-index (GRIN) lens forms a parallel beam that passes the photorefractive material and that is coupled by another GRIN lens into the throughput fiber. However, inside the photorefractive material there are four holograms that are Bragg-matched to four of the channel wavelengths λ_i, λ_j, λ_k, and λ_l. The diffracted waves should have a spherical shape. At the focal point a fiber end is positioned that collects the dropped light. Since the four waves are diffracted from different holograms, the focal spots can be separated and the light is coupled into four independent fibers. Compared to other devices this approach has the advantage that no additional optics in front of these four fibers is required. This makes it possible to fabricate more compact, simpler and cheaper devices.

For fabrication of devices like that shown in Fig. 9 one problem typically appears that can be also solved by IR recording: it is the alignment issue. The hologram must not only be properly recorded, but also the photorefractive material and all the fibers must be precisely positioned in order to keep the insertion loss and the drop losses small. For IR recording this can be achieved easily: the device can be built first and after fixing the fibers, e.g. with glue, recording can be performed. Two-step writing requires the presence of sensitizing light. The IR recording light is then provided through the fibers, which ensures for the final operation perfect alignment of the hologram and of the fibers. Although this method has not been experimentally demonstrated so far, it seems to be straightforward that next-generation components may use this technology.

6 Conclusions

Many photorefractive crystals which are insensitive in the IR spectral region may be sensitized for IR recording by two-step processes. Nondestructive readout of the holograms recorded by two-step processes is possible. In contrast to other methods for hologram stabilization, e.g. thermal fixing, the versatility of desired optical erasure is maintained. The lifetime of the holograms can approach years in materials like $LiNbO_3$ if the doping level is optimized and if the crystals are dehydrated. Direct IR recording with light of the operational wavelength has two practical advantages: (1) Not only gratings, but also more sophisticated components can be fabricated. A wavelength filter that focuses the diffracted light into a fiber is one example. (2) The holograms can be recorded in the final device. This simplifies assembling and adjustment of the components.

Acknowledgments

The authors are indebted to A. Gerwens, H. Hesse, L. Holtmann, J. Imbrock, F. Jermann, D. Kip, M. Müller, I. Nee, M. Ye, M. Simon, S. Wevering and H. Vormann for valuable help and discussions. Financial support of the Deutsche Forschungsgemeinschaft and of the Volkswagen-Stiftung is gratefully acknowledged.

References

1. P. Günter, J.-P. Huignard, (Eds.): *Topics in Applied Physics: Photorefractive Materials and Their Applications II*, Topics Appl. Phys. **62**, (Springer, Berlin, Heidelberg 1989)
2. K. Buse, E. Krätzig: Inorganic Photorefractive Materials, in H. Coufal, D. Psaltis, G.Sincerbox (Eds.): *Holographic Storage* (Springer, Berlin, Heidelberg, 2000)
3. D. von der Linde, A. M. Glass, K. F. Rodgers: Multiphoton photorefractive processes for optical storage in $LiNbO_3$, Appl. Phys. Lett. **25**, 155 (1974)
4. D. von der Linde, A. M. Glass, K. F. Rodgers: high-sensitivity optical recording in KTN by two-photon absorption, Appl. Phys. Lett. **26**, 22 (1975)
5. H. Vormann, E. Krätzig: Two-step excitation in $LiTaO_3$:Fe for optical data storage, Solid State Commun. **49**, 843 (1984)
6. Y. Ming, E. Krätzig, R. Orlowski: Photorefractive effects in $LiNbO_3$:Cr induced by two-step excitation, Phys. Status Solidi A **92**, 221 (1985)
7. A. Motes, J. J. Kim: Intensity-dependent absorption coefficient in photorefractive $BaTiO_3$ crystals, J. Opt. Soc. Am. B **4**, 1379 (1987)

8. G. A. Brost, R. A. Motes, J. R. Rotgé: Intensity-dependent absorption and photorefractive effects in barium titanate, J. Opt. Soc. Am. B **5**, 1879 (1988)
9. L. Holtmann: A model for the nonlinear photoconductivity of $BaTiO_3$, Phys. Status Solidi A **113**, K89 (1989)
10. L. Holtmann, K. Buse, G. Kuper, A. Groll, H. Hesse, E. Krätzig, Photoconductivity and light-induced absorption in $KNbO_3$:Fe, Appl. Phys. A **53**, 81 (1991)
11. K. Buse, E. Krätzig: Light-Induced Charge Transport in Photorefractive Crystals, in F. Yu, S. Yin: *Photorefractive Optics: Materials, Properties and Applications* (Academic Press, New York 2000)
12. K. Buse, E. Krätzig: Three-valence charge-transport model for explanation of the photorefractive effect, Appl. Phys. B **61**, 27 (1995)
13. K. Buse, A. Adibi, D. Psaltis: Non-volatile holographic storage in doubly doped lithium niobate crystals, Nature **393**, 665 (1998)
14. K. Buse, L. Holtmann, E. Krätzig: Activation of $BaTiO_3$ for infrared holographic recording, Opt. Commun. **85**, 183 (1991)
15. A. Gerwens, M. Simon, K. Buse, E. Krätzig: Activation of cerium-doped strontium barium niobate for infrared holographic recording, Opt. Commun. **135**, 347 (1997)
16. A. Kamshilin, M. P. Petrov: Infrared quenching of the photoconductivity and holographic data storage in $Bi_{12}SiO_{20}$, Sov. Solid State Phys. **23**, 3110 (1981)
17. S. G. Odoulov, K. V. Shcherbin, A. N. Shumeljuk: Photorefractive recording in BTO in the near infrared, J. Opt. Soc. Am. B **11**, 1780 (1994)
18. S. G. Odoulov, A. N. Shumelyuk, U. Hellwig, R. A. Rupp, A. A. Grabar, I. M. Stoyka: Photorefraction in tin hypothiodiphosphate in the near infrared, J. Opt. Soc. Am. B **13**, 2352 (1996)
19. P. Pogany, H. J. Eichler, M. Hage Ali: Two-wave mixing gain enhancement in photorefractive CdZnTe:V by optically stimulated electron-hole resonance, J. Opt. Soc. Am. B **15**, 2716 (1998)
20. K. Shcherbin, F. Ramaz, B. Farid, B. Briat, H.-J. von Bardesleben: Photoinduced charge transfer processes in photorefractive CdTe:Ge, OSA TOPS **27**, 54 (1999)
21. D. von der Linde, A. M. Glass: Photorefractive effects for reversible holographic storage of information, Appl. Phys. **8**, 85 (1975)
22. F. Jermann, J. Otten: Light-induced charge transport in $LiNbO_3$:Fe at high light intensities, J. Opt. Soc. Am. B **10**, 2085 (1993)
23. M. Simon, F. Jermann, E. Krätzig: Intrinsic photorefractive centers in $LiNbO_3$:Fe, Appl. Phys. B **61**, 89 (1995)
24. K. Buse, F. Jermann, E. Krätzig: Infrared holographic recording in $LiNbO_3$:Cu, Appl. Phys. A **58**, 191 (1994)
25. K. Buse, F. Jermann, E. Krätzig: Infrared holographic recording in $LiNbO_3$:Fe and $LiNbO_3$:Cu, Opt. Mater. **4**, 237 (1995)
26. J. Imbrock, S. Wevering, K. Buse, E. Krätzig: Nonvolatile holographic storage in photorefractive lithium tantalate crystals with laser pulses, J. Opt. Soc. Am. B **16**, 1302 (1999)
27. A. M. Glass, D. von der Linde, T. J. Negran: High-voltage bulk photovoltaic effect and the photorefractive process in $LiNbO_3$, Appl. Phys. Lett. **25**, 233 (1974)
28. Y. S. Bai, R. Kachru: Nonvolatile Holographic Storage with two-step recording in lithium niobate using cw lasers, Phys. Rev. Lett. **78**, 2944 (1997)

29. H. Guenther, G. Wittmann, R. M. Macfarlane, R. R. Neurgaonkar: Intensity dependence and white-light gating of two-color photorefractive gratings in $LiNbO_3$, Opt. Lett. **22**, 1305 (1997)
30. J. Imbrock, D. Kip, E. Krätzig: Nonvolatile holographic storage in iron-doped lithium tantalate with continuous wave laser light, Opt. Lett. **24**, 1302 (1999)
31. M. Horowitz, B. Fischer, Y. Barad, Y. Silberberg: Photorefractive effect in a $BaTiO_3$ crystal in the 1.5 µm wavelength regime by two-photon absorption, Opt. Lett. **21**, 1120 (1996)
32. K. Oba, P.-C. Sun, Y. Fainman: Nonvolatile photorefractive spectral holography, Opt. Lett. **23**, 915 (1998)
33. V. Leyva, G. A. Rakuljic, B. O'Conner: Narrow bandwidth volume holographic optical filter operating at the Kr transition at 1547.82 nm, Appl. Phys. Lett. **65**, 1079 (1994)
34. R. Müller, M. T. Santos, L. Arizmendi, J. M. Cabrera: A narrow-band interference filter with photorefractive $LiNbO_3$, J. Phys. D: Appl. Phys. **27**, 241 (1994)
35. S. Breer, K. Buse: Wavelength demultiplexing with volume phase holograms in photorefractive lithium niobate, Appl. Phys. B **66**, 339 (1998)
36. S. Breer, H. Vogt, I. Nee, K. Buse: Low-crosstalk WDM by Bragg diffraction from thermally fixed reflection holograms in lithium niobate, Electron. Lett. **34**, 2419 (1999)
37. J. J. Amodei, D. L. Staebler: Holographic pattern fixing in electro-optic crystals, Appl. Phys. Lett. **18**, 540 (1971)
38. K. Buse, S. Breer, K. Peithmann, S. Kapphan, M. Gao, E. Krätzig: Origin of thermal fixing in photorefractive lithium niobate crystals, Phys. Rev. B **56**, 1225 (1997)
39. L. Arizmendi, E. M. Miguel-Sanz, M. Carrascosa: Lifetimes of thermally fixed holograms in $LiNbO_3$:Fe crystals, Opt. Lett. **23**, 960 (1998)
40. H. Vormann, G. Weber, S. Kapphan, E. Krätzig, Hydrogen as origin of thermal fixing in $LiNbO_3$:Fe, Solid State Commun. **40**, 543 (1981)
41. I. Nee, K. Buse, F. Havermeyer, R. A. Rupp, M. Fally, R. P. May: Neutron diffraction from thermally fixed gratings in photorefractive lithium niobate crystals, Phys. Rev. B **60**, R9896 (1999)
42. H. C. Külich: A new approach to read volume holograms at different wavelengths, Opt. Commun. **64**, 407 (1987)
43. I. Nee, M. Müller, K. Buse, E. Krätzig: Role of iron in lithium niobate crystals for the dark storage time of holograms, J. Appl. Phys. **88**, 4282 (2000)
44. Y. P. Yang, I. Nee, K. Buse, D. Psaltis: Ionic electronic dark decay of holograms in $LiNbO_3$ crystals, Appl. Phys. Lett. **78**, 4076 (2001)
45. K. Buse: Light-induced charge transport processes in photorefractive crystals II: Materials, Appl. Phys. B **64**, 391 (1997)
46. S. Brülisauer, D. Fluck, P. Günter, L. Beckers, C. Buchal: Photorefractive effect in proton-implanted Fe-doped $KNbO_3$ waveguides at telecommunication wavelengths, J. Opt. Soc. Am. B **13**, 2544 (1996)

Gated Optical Recording for Nonvolatile Holography in Photorefractive Materials

Lambertus Hesselink and Sergei S. Orlov

Stanford University
CIS-X Building, Stanford, CA 94305-4075
bert@kaos.stanford.edu

Abstract. Hologram recording in photorefractive media is classically performed in the blue-green spectral range and suffers from optical erasure under continuous readout, unless a thermal fixing technique is applied. Gated optical holographic recording allows us to obtain high-diffraction efficiencies in the near InfraRed (IR) jointly with long-term storage. In particular, the roles of stoichiometry and dopant reduction state in $LiNbO_3$ in the optimization of sensitivity is analyzed, in order to perform digital information storage in such materials via the two-photon technique.

1 Introduction

Illumination of a spatially visible wavelength light pattern in a photorefractive medium such as $LiNbO_3$ – for example derived from the superposition of two mutually coherent plane waves – stimulates a charge distribution that gives rise to a corresponding change in the index of refraction of the medium through the electro-optic effect. Although several undoped ferroelectric media exhibit photorefractive response, the strength of the effect can be improved by orders of magnitude by adding selected dopants to the medium during crystal growth. For example, iron (Fe) is a well-known dopant in congruent $LiNbO_3$ (when $LiNbO_3$ is grown by melting Li_2CO_3 and Nb_2O_5, the congruently melted composition is 48.6 mol% Li_2O, as opposed to 50 mol% for stoichiometric $LiNbO_3$). Iron is present in the crystal as Fe^{2+} and Fe^{3+}, in a ratio determined by post-crystal-growth annealing in an oxidizing or reducing atmosphere. The Fe^{2+} acts as a donor ion, which upon absorption of a photon produces an electron and Fe^{3+}, which acts as an acceptor for free carriers. The free electrons move in the crystal from regions of intense illumination to darker areas under diffusion, drift or photovoltaic forces. Acceptors capture the free electrons, reducing Fe^{3+} to Fe^{2+}. Through repeated steps, electrons tend to be trapped by Fe^{3+} atoms in regions where the light intensity is low. The charge redistribution gives rise to a spatially varying electric space-charge field that changes the index of refraction of the material locally through the electro-optic effect and represents the holographic information to be stored. Upon readout of the hologram, the stored information is retrieved, but also erased because the medium remains photosensitive. The readout beam, which usually is a uniform plane wave, excites free electrons again, and redistributes the charges uniformly throughout the recording medium.

P. Boffi, P. Piccinin, M. C. Ubaldi (Eds.): Infrared Holography for Optical Communications, Topics Appl. Phys. **86**, 41–57 (2003)

Photorefractive recording materials therefore require a fixing process to reduce or eliminate their sensitivity to visible recording wavelengths as they are being read out. This problem appears to be intrinsic to holographic media because the materials have to have a linear response to exposure in order to record superimposed (multiplexed) volume holograms with a minimum of scatter. To avoid this problem, a nonlinear recording medium with a steep threshold response to illumination is required, so that nondestructive readout can occur at much lower power levels than used during recording – as is the case for media used in Digital Versatile Disc-Rewritable (DVD-RW) and Compact Disc-Rewritable (CD-RW) devices. These media exhibit nonlinear temperature-induced behavior through phase changes, where the crystalline structure changes into an amorphous one and vice versa, with a companying change in reflectivity. In photorefractive media temperature-dependent nonlinear effects are difficult to achieve through illumination, as volumetric media cannot be significantly heated by light absorption from small lasers. They must be largely transparent for volumetric recording. Temperature-dependent effects are used for fixing holographic gratings in $LiNbO_3$, however, but only by resorting to post-recording annealing at temperatures around 150 °C of the bulk materials having substantial concentrations of ionic dopants. For example, H-ions are not light sensitive, but they become mobile at higher temperatures and collectively can mimic the electronic space-charge field recorded at room temperatures, where they are stable. Optical erasure upon readout by a uniform laser beam erases the electronic space-charge field, revealing the ionic grating. Strong and permanent gratings can thus be recorded in $LiNbO_3$ crystals using thermal fixing techniques, but this is not a practical procedure. Alternatively, the photorefractive medium can be made significantly less sensitive to the readout wavelength through an optical gating approach.

Optically gated recording focuses on using an additional light source during the recording process that is not present during readout [1]. The gating light source can be incandescent and does not have to be coherent. The material is sensitive to the writing wavelength only in the presence of gating or bias light, while it is insensitive to readout light by itself. This optically induced light sensitivity is potentially a very attractive mechanism for data recording and readout, as it allows long-term storage and virtually nondestructive retrieval. Additional advantages of the two-photon recording scheme include low absorption of the reading light (which produces stronger signals for the same index modulation) and virtual absence of optical damage during hologram readout. Furthermore, optically induced sensitivity is increased in the near-IR region of the spectrum, where most photorefractive materials are insensitive to single-photon recording.

In early work, pulsed lasers were used to achieve large intensities to take advantage of optical nonlinearities. Transition metals were doped into a variety of host materials, including $LiNbO_3$, KTN and $LiTaO_3$. Experiments showed that by using a fundamental harmonic (1.064 μm) of a YAG-laser for writing the data and its second harmonic (532 nm) for gating, long erasure times could be obtained, but the required near-IR intensities were on the order of $10^8\,W/cm^2$ [1]. Alternatively, Mn- and Fe- co-doped photochromic congruent $LiNbO_3$ crystals and rare-earth (e.g. Pr) ions doped into $LiNbO_3$ and SBN materials have been used [2,3,4] for two-photon recording.

In recent years, substantial advances in two-photon recording with near-infrared CW lasers have been made using near-stoichiometric doped and undoped lithium niobate material [5,6,7]. The key idea for optimization was to modify the host composition and the reduction state of the crystalline material [5,6]. It has been demonstrated that even for moderate CW intensities of the recording light ($5\,W/cm^2$ or less) the two-photon sensitivities of near-stoichiometric reduced lithium niobate material can be as high as the non-gated recording sensitivity of Fe-doped $LiNbO_3$ crystals illuminated by only green light [5,6]. Digital hologram storage has also been successfully demonstrated using these improved gated two-photon materials [7].

To characterize the sensitivity of a medium, several measures have been developed that generally relate the measured diffraction efficiency or the index of refraction change under illumination to theincident light intensity. The recording photosensitivity is typically defined a follows:

$$S(I_{\mathrm{g}}) = \frac{\partial\sqrt{\eta}}{\partial t}\frac{1}{I_{\mathrm{w}}L} \tag{1}$$

where η is the diffraction efficiency, L is the thickness of the medium, I_{w} is the recording intensity and I_{g} is the gating intensity. The typical sensitivity of Fe-doped $LiNbO_3$ in the green is from 0.02 to 0.10 cm/J, depending on the total doping level and the reduction state of the Fe-dopant. The best $LiNbO_3$-based two-photon materials now have similar sensitivities in the near-IR.

2 Gated Recording in Undoped Stoichiometric Lithium Niobate

In order to achieve optically gated recording, there must be present:

- deep traps that are partially filled with electrons to generate excited carriers during gating and recording of the hologram
- shallow levels to trap photogenerated electrons, with sufficiently long lifetimes (hundreds of milliseconds) to absorb light at the writing wavelength.

Efficient two-photon gated recording can be realized using intrinsic point defects of lithium niobate. Lithium niobate crystals are in general not stoichiometric, as the chemical formula is given by $Li_{1-x}Nb_{1+x/5}O_3$ (and $x = 0.028$ for congruently melting material). Intrinsic point defects include Nb_{Li} (that is, a niobium ion on the Li-site) antisite defects: $Nb_{Li}V_{Nb}$ and $Nb_{Li}Nb_{Nb}$ [8]. The latter represents a two-electron trap, or, when filled with a pair of electrons with opposite spins, a diamagnetic bipolaron $Nb^{4+}Nb^{4+}$. Illumination with blue-green light dissociates the bipolarons, and these electrons enter the

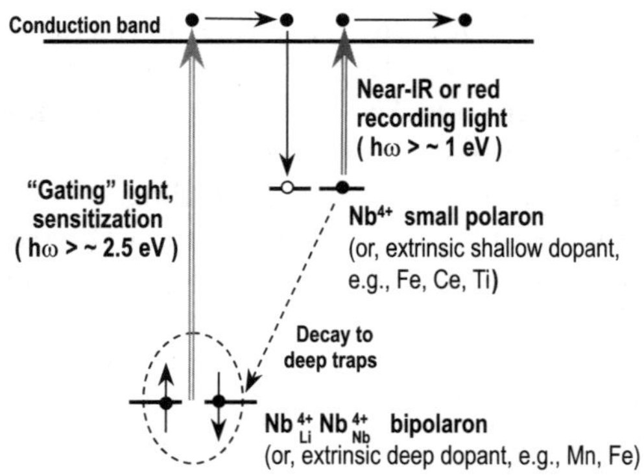

Fig. 1. Mechanism of optically gated recording in lithium niobate

conduction band and become self-trapped on the Nb lattice sites, forming Nb^{4+} small polarons [9] (Fig. 1). Small polaron levels can be viewed as shallow traps with a limited electron lifetime. For small polarons, the absorption due to optical excitation (hopping between different sites) is very broadband and peaks in the near-infrared region of the spectrum at about 780 nm, which is twice the binding energy of polarons (0.8 eV). Photoexcitation by recording with near-infrared light and subsequent carrier redistribution and migration creates the index hologram. The lifetime of the excited Nb^{4+} states depends on the density, capture cross-section and the reduction state of the deep defects, as well as the temperature and the intensity of the recording IR light. When the gating light is turned off, the carriers eventually become trapped in the deep traps and the material becomes essentially transparent to the near-IR readout radiation.

Optimization of the undoped crystalline material can be achieved through changing its stoichiometry and the reduction state[5,6] (Fig. 2). The gated photosensitivity of the material to the near-IR light is proportional to the density of excited small polarons, and, therefore, is proportional to their recombination lifetime. High defect densities, such as occur in congruent lithium niobate (48.6 mol% Li_2O) lead to substantial lattice disorder and fast non-radiative, phonon-assisted decay of Nb^{4+} polaron levels into deep traps and their recombination to bound bipolarons [8,10]. As the material becomes closer to stoichiometric and the defect density decreases, the two-photon sensitivity S exhibits a sharp, two-order-of-magnitude increase with increased Li/Nb ratio up to 49.6 mol% and then saturates (Fig. 2). The lifetime of the upper Nb^{4+} levels in lithium niobate at room temperature changes from $\leq$ 40 ms in as-grown composition [2] to several seconds in lightly reduced near-stoichiometric (with $[Li_2O] = 49.65$ mol%) lithium niobate (Fig. 3).

Post-growth reduction of near-stoichiometric crystals produces more filled bipolaron sites, which provide efficient absorption for the gating light of the blue-green region of the spectrum. Also, reduction converts the contaminant

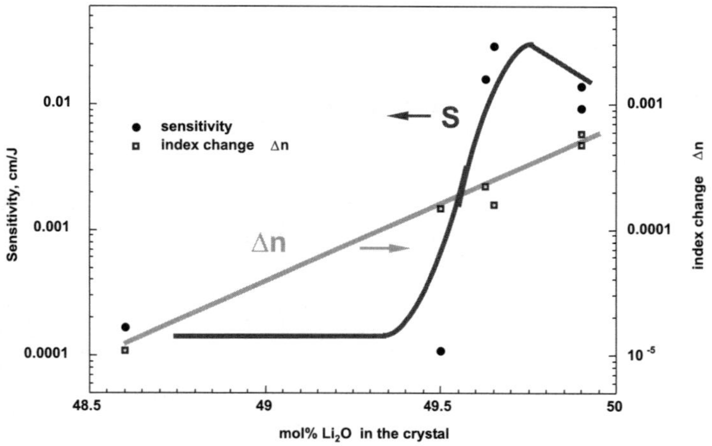

Fig. 2. Gated photosensitivity and index change versus material composition (Li content). All crystals are lightly reduced. Gating intensity is $0.25\,\mathrm{W/cm^2}$ at 457.9 nm

acceptor ions into donors (for example, Fe^{3+} into Fe^{2+}), which decreases the small polaron capture cross-section and increases their lifetime and, therefore, the material sensitivity under similar gating conditions. The lifetime of small polarons in optimally reduced crystals can be as long as several seconds and depends on the degree of reduction. More excessive reduction, however, leads to very long polaron lifetimes (tens of seconds) that adversely affect recording, in that the holograms exhibit a substantial dark decay right after writing. Optimally reduced samples, however, exhibit both high sensitivity and high gating ratio (≥ 100), allowing both fast recording and prolonged nondestructive readout (Fig. 4). The gating ratio is defined here to be the ratio of the readout time constants with gating light absent and present.

Reduction of lithium niobate at high temperatures (900 to 1000 °C) involves the migration of lithium vacancies from the bulk to the surface of the material (with corresponding back-transport of free electrons for charge compensation). Surface reaction at the processing temperature involves the consumption of nonstoichiometric cells and the generation of free electrons [11]:

$$2(\mathrm{Nb}_{\mathrm{Li}}^{5+}\mathrm{V}_{\mathrm{Nb}}^{5-})^{1-}\mathrm{O}_3 \rightarrow 3/2\mathrm{O}_2 + (\mathrm{Nb}_{\mathrm{Li}}^{5+}\mathrm{Nb}_{\mathrm{Nb}}^{5+})^{4+}\mathrm{O}_3 + \mathrm{V}_{\mathrm{LiNbO_3}} + 6e^- \,. \qquad (2)$$

At lower temperatures free electrons are trapped on the defect sites forming bipolarons, while local charge compensation is still maintained by a change of the density of lithium vacancies:

$$(\mathrm{Nb}_{\mathrm{Li}}^{5+}\mathrm{Nb}_{\mathrm{Nb}}^{5+})^{4+} + 2e^- \rightarrow (\mathrm{Nb}_{\mathrm{Li}}^{4+}\mathrm{Nb}_{\mathrm{Nb}}^{4+})^{2+} \,. \qquad (3)$$

A substantial increase in the blue-green absorption is seen in the reduced materials (Fig. 5). After reduction the two-photon photosensitivity of near-stoichiometric material is about 20 times higher than in as-grown

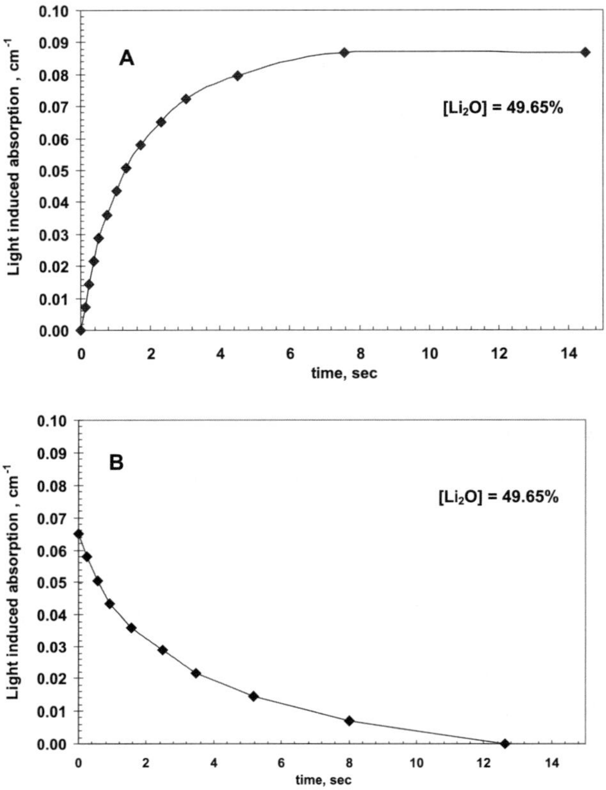

Fig. 3. Build-up **(a)** and dark decay **(b)** of light-induced absorption at $\lambda_{\mathrm{polar}} =$ 670 nm) in a stoichiometric lightly reduced lithium niobate sample. Gating intensity is $0.5\,\mathrm{W/cm^2}$ at 532 nm

materials [5,6]. The sensitivity saturates for undoped materials (or even decreases) for near-stoichiometric compositions with $[\mathrm{Li_2O}] \geq 49.65$ mol%. Since the density of the defects ($\mathrm{Nb_{Li}Nb_{Nb}}$) decreases as the composition becomes closer to exactly stoichiometric, the absorption of the gating light (and, hence, the two-photon sensitivity) becomes somewhat smaller for more stoichiometric reduced crystals (see Figs. 5 and 2). The saturation index change (i.e. dynamic range), however, continues to grow with decreasing defect density.

Sensitivity measurements are usually performed in the holographic transmission recording geometry (30° between the writing IR beams in air) with extraordinarily polarized infrared 800 nm Ti:sapphire laser light for recording and an expanded argon ion laser beam for gating. For a constant nearinfrared writing intensity the dependence of the two-photon photosensitivity S on the gating intensity $\mathrm{I_g}$ (Fig. 6) can be approximated as follows:

$$S(I_{\mathrm{g}}) = S_{\max} \frac{I_{\mathrm{g}}}{I_{\mathrm{g}} + I_{\mathrm{S},1/2}}\,, \tag{4}$$

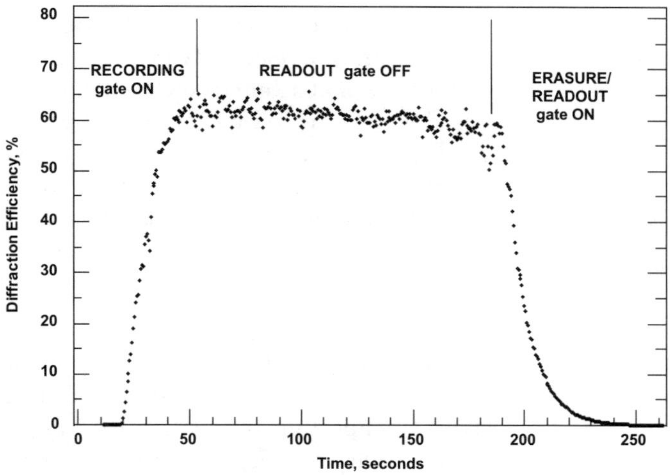

Fig. 4. Gated recording, nondestructive readout (with gating light off) and erasure (with gate on) in an undoped lightly reduced crystal (Li_2O content is 49.6 mol%; IR intensity is 6 W/ cm^2 and gating intensity is 0.25 W/ cm^2 at 457.9 nm)

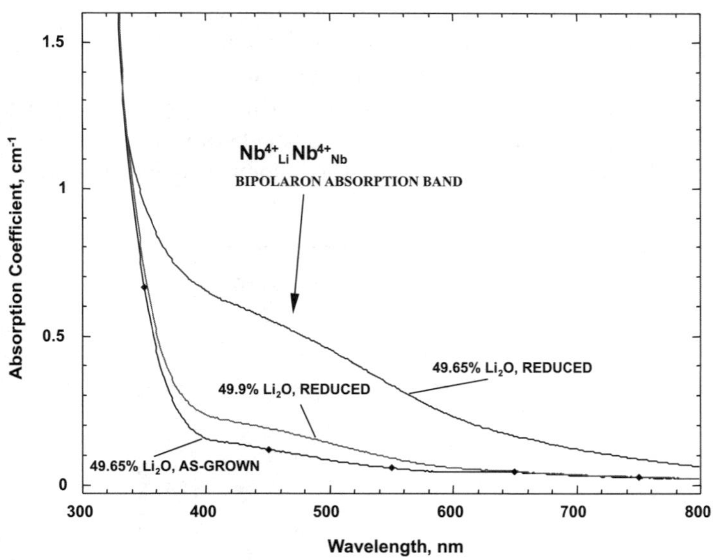

Fig. 5. Bipolaron absorption induced by the reduction of lithium niobate

where $I_{S,1/2}$ is the gating intensity at which the sensitivity attains a value of 1/2 of that at saturation and S_{max} is the saturation sensitivity for high gating powers. The saturation in infrared sensitivity is attributed to the exhaustion of the bipolarons at high gating intensities. Increased gating intensity also leads to erasure and reduction of the fringe visibility of the grating recorded with near-infrared light. As a result the saturation index change Δn (and corresponding M♯, as they are linearly related to each other) falls off for high gating powers:

$$\Delta n(I_g) = \Delta n_{max} \frac{I_{\Delta n,1/2}}{I_g + I_{\Delta n,1/2}}, \quad (5)$$

where $I_{\Delta n,1/2}$ represents the value of the gating intensity at which the holographic index change Δn decreases by a factor of two of its maximum value, which occurs at zero gating intensity. It is important to mention that $I_{S,1/2}$ and $I_{\Delta n,1/2}$ are not necessarily equal and that both depend strongly on the gating light wavelength.

Both the sensitivity and the index change are enhanced in crystals with increased bipolaron concentration (Figs. 2 and 5). In the presence of the gating light, the induced absorption at the recording near-IR wavelength (800 nm) is of the order of 0.1 cm^{-1} (Fig. 3). Without gating, the erasure upon readout and the IR one-photon sensitivity is determined essentially by the absorption tail of deep impurities in the near-IR. The sensitivity of the material to non-gated recording in the near IR is at least 50 to 1000 times less, which allows readout of the hologram thousands of times without substantial erasure and optically induced index damage. Dark decay of the hologram is determined by

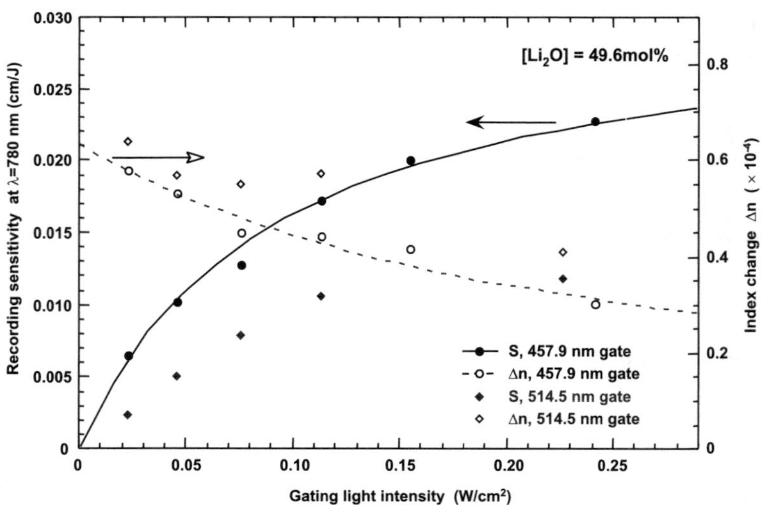

Fig. 6. Two-photon photosensitivity and saturation index change versus 457.9 nm and 514.5 nm gating light intensity in reduced lithium niobate crystal. Writing intensity is 10 W/cm^2

thermal excitation of electrons from deep bipolaron sites into the conduction band. Depending on the reduction state of the material the room-temperature lifetime can vary from several weeks to several months, or longer. The photoconductivity of reduced near-stoichiometric lithium niobate crystals also exhibits sublinear behavior, as deduced from the erasure rate of holograms versus gating light intensity dependence (Fig. 7). For a 457.9 nm gate, the photoconductivity of a crystal containing 49.6 mol% Li_2O is proportional to $\approx I_g^{0.75}$ [5,6].

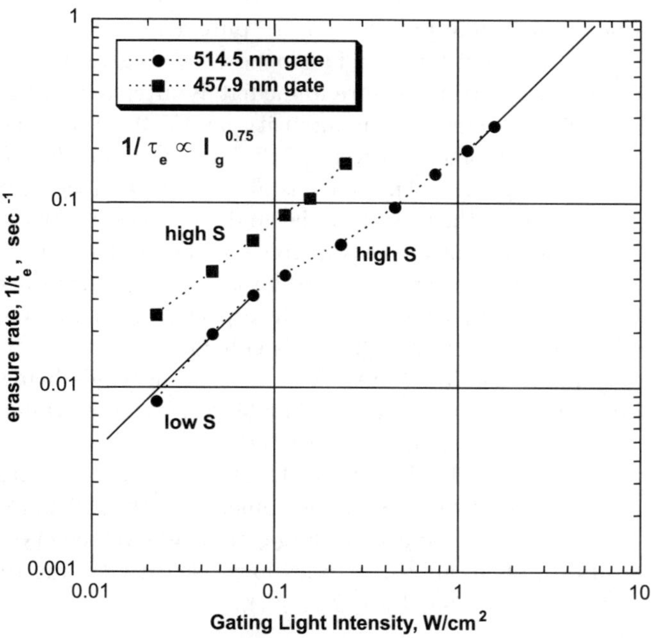

Fig. 7. Hologram erasure rate versus erasing light intensity in a 49.6 mol% reduced nominally undoped lithium niobate crystal

3 Doped Stoichiometric Lithium Niobate

Extrinsic dopants can play essentially the same role as deep bipolarons provided that their energy levels are sufficiently deep in the forbidden gap of the crystal (Fig. 8). Normally, the ionization absorption for dopants in $LiNbO_3$ is quite broadband, and, therefore, in order to achieve high gating efficiency it is important to separate the absorption peaks for deep donors and metastable polarons ($\approx$ 780 nm) as far as possible. Mn^{2+}, Fe^{2+} and Ce^{3+} are appropriate deep donors because their energy levels lie below that of the bipolarons in the forbidden band gap [8], and, therefore, one should expect a smaller photoionization cross-section for near-IR light for these donors compared to bipolarons. Weakly doped crystals exhibit higher gating efficiency of the two-photon recording process and, therefore, longer hologram lifetime upon continuous readout [6]. Reduction converts the majority of the extrinsic acceptors (such as Fe^{3+} or Mn^{3+} ions) into donors (Fe^{2+} and Mn^{2+}). For 0.02 mol% Mn-doped samples, at least 90 to 95% of the Mn ions need to be converted (via reduction) into the Mn^{2+} valence state before any appreciable two-photon response can be measured [6]. The optimum reduction for doped crystals requires stronger reducing conditions compared to those for undoped crystals. Typical processing conditions involve temperatures of about 950 °C to 1000 °C and duration of about 30 min in a flow of dry pure argon. As in undoped material, stronger reduction of Mn- and Fe- doped crystals gives rise to high dark conductivity and fast grating erasure immediately after recording.

Figure 9 shows the sensitivity versus gating intensity (at 457.9 nm) dependence in reduced Fe-doped stoichiometric crystals [12] ($[Li_2O] = 49.9$ mol% and $[Fe] = 0.01$ wt%). The two-photon sensitivity is about 30% and the dynamic range is about 60% higher than that of the undoped crystals (with $[Li_2O] = 49.6$ mol%). An improvement in the sensitivity at 488 and 457.9 nm (compared with that at 514 nm) is more pronounced for the Fe-doped crystals than for the undoped crystals. Similar behavior with respect to gating wavelength was also found in Mn-doped crystals [6]. This enhanced sensitivity in doped lithium niobate is due to the Mn^{2+} and Fe^{2+} absorption bands being blue-shifted as compared to the bipolaron band. The largest dynamic range is found in lightly reduced undoped material with the smallest defect density ($\Delta n \approx 0.6 \times 10^{-3}$ for undoped $LiNbO_3$ with $[Li_2O] = 49.9$ mol%). The undoped crystals of a composition near exact stoichiometry (with 49.9 mol% Li_2O content) have a dynamic range about twice larger than a Fe-doped crystal of the same composition but about three times lower sensitivity due to lower absorption of the gating light (see Figs. 2 and 5).

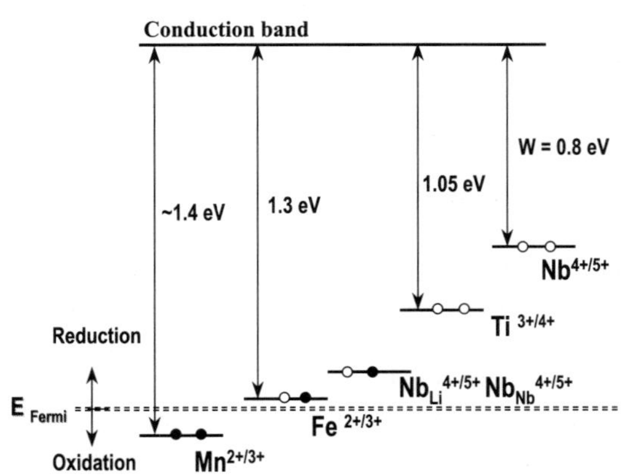

Fig. 8. Energy levels of different impurities and intrinsic defect centers in $LiNbO_3$

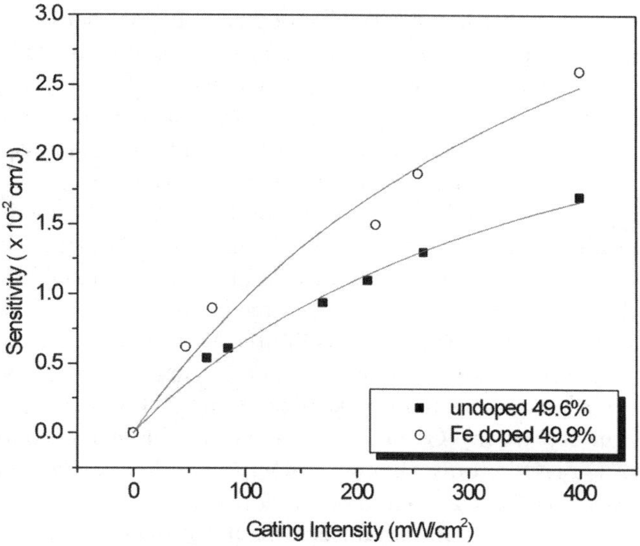

Fig. 9. Two-photon sensitivity vs. 457.9 nm gating intensity of reduced stoichiometric (49.9 mol% Li_2O) Fe-doped and reduced undoped (49.6 mol% Li_2O) $LiNbO_3$. Writing intensity is $10\,W/cm^2$

Comparison of the Fe-doped and the undoped crystals also indicates that the erasure upon continuous IR readout is approximately 3 times slower in a Fe-doped sample compared to that in an undoped crystal with 49.6 mol% Li_2O content and is comparable to that in the undoped 49.9 mol% sample. The latter, however, has lower (×3) sensitivity than the Fe-doped samples. The two-photon sensitivity decreases (by a factor of 2) both in Mn- and Fe-doped samples when the dopant content is increased from 0.01 mol% to 0.03 mol%, which indicates that an optimal doping concentration exists for a high sensitivity and decreased IR erasure rate [13]. With increasing concentration of acceptors in the crystal, recombination [13] from the small polarons into the deep traps (i.e. Fe^{3+} or Mn^{3+}) becomes more pronounced, which decreases the lifetime of the small polaron concentration, and thereby decreases the two-photon sensitivity.

4 Material Preparation and Characterization

At Stanford University we typically grow lithium niobate crystals in air by the Czochralski technique using radio-frequency (rf) coupling. The crystals are pulled along the hexagonal *c*-axis at the rate of 0.5 to 2 mm/h and a rotation rate of 15 to 20 rpm. To obtain a single domain, the crystals are poled by applying an electric field above the Curie temperature. The Nb_2O_5 and Li_2CO_3 used for preparing the charges are commonly obtained from AESAR and have a purity of 99.99%. Crystals of $LiNbO_3$ are typically grown from a congruent composition with a 48.6 mol% of Li_2O. In our work the Li_2O mol% in the melt varied from 48.6% to 58%, which corresponds to 48.6% to 49.9% in the crystal [14]. The Curie temperature increases with increasing Li content [14]. The crystals are oriented along the *c*-axis using the Laue method and cut and polished for optical measurements. Reduction of crystal samples is usually performed in a high-temperature tube furnace in a flow of dry ultra-high-purity argon gas. Special care needs to be taken to avoid oxygen contamination of the processing chamber. The minimum flow used is $10\,cm^3/s$ and a long thin exhaust tubing is used to avoid back-diffusion of atmospheric oxygen and water vapor into the processing chamber. The oxygen content during processing is under 10 ppm and is limited by the purity of the argon gas used. Optimum reduction steps for undoped near-stoichiometric material usually require a processing temperature of about 900 °C for a duration of 15 to 30 min (in dry argon). Crystals nearer the stoichiometric composition (e.g. 49.9 mol% Li_2O) require lighter reducing conditions. Doped crystals, if reduced under conditions optimal for undoped materials, usually exhibit little or almost no two-photon response. Optimal processing conditions for Mn- and Fe-doped crystals involve temperatures of about 950 °C to 1000 °C and durations of about 30 min in a flow of dry pure argon. As in undoped material, stronger reduction of Mn- and Fe-doped crystals gives rise to high dark conductivity and fast grating erasure immediately after recording. Undoped crystals, if processed in conditions optimal for doped samples, become overreduced and exhibit high dark conductivity and fast erasure of recorded holograms. Reduction in vacuum (of undoped or doped crystals) may give similar results for the two-photon sensitivity and reduction state of the material, but, unfortunately, also leads to optical surface degradation (e.g. extra roughness and darkening) and, therefore, was not used in our work. The Li/Nb ratios of the off-congruent compositions are characterized using infrared spectroscopy and the measurements of the second-harmonic generation phase-matching temperature. $LiNbO_3$ grown by the Czochralski method contains protons in the form of OH^- ions. The line shape of the OH^- bonds stretching vibration absorption in the IR ($\approx 2.87\,\mu m$) was used to determine the Li/Nb stoichiometry [15]. The IR absorption spectra are recorded using a Hitachi U-4001 spectrophotometer. Lithium niobate crystals exhibit a pronounced structure in the O–H stretching vibration absorption in the IR region, with the intensity of the peaks varying with the Li/Nb ratio in the crystals [6]. With increasing Li deficiency the intensity of the two highest energy transitions increases (3499 and $3490\,cm^{-1}$), while that of the two others decreases (3481 and $3470\,cm^{-1}$). The lowest energy component ($3466\,cm^{-1}$) is very sensitive to the changing Li/Nb composition and disappears in the congruent composition. The structure associated with the IR peaks is due to five different proton sites in the oxygen planes of $LiNbO_3$, each having different cationic environments [15]. Because the birefringence of $LiNbO_3$ changes with its composition the Li/Nb ratio of the crystals can also be verified using phase-matching temperature measurements [16]. We normally perform phase-matching measurements using a 90° noncritical phase-matching configuration with a Nd:YAG laser operating at 1064 nm. The compositions of $LiNbO_3$ samples are determined using the relation between phase matching temperatures (T_{PM}) and the mol% Li_2O in $LiNbO_3$ [16]. There is an increase in the phase-matching temperature with an increase in the Li/Nb concentration; a nearly stoichiometric crystal has the highest phase-matching temperature of 227 °C. For example, a nominal composition of the starting melt of 54 mol% Li_2O and 46 mol% Nb_2O_5 leads to the crystal composition of $Li_{0.9865}Nb_{1.0027}O_3$, i.e. 49.7 mol% Li_2O and 50.3 mol% Nb_2O_5 in the crystal.

5 Digital Information Storage Experiment in Two-Photon Photorefractive Material

As described above, in two-photon gated recording one wavelength is used to sensitize the recording medium, while the other wavelength is used to write the information. Gating light populates an intermediate state in the bandgap of the material, thus making the medium sensitive to long writing wavelengths. In the presence of gating the writing light excites charge carriers from that state into the conduction band, where they migrate and recombine with deep traps. When the gating is switched off, the media becomes no longer sensitive for the writing light, allowing nondestructive reading. This process eliminates the need for any postprocessing and simplifies system design. Other advantages of two-color recording are significant: there is no light absorption during readout at the longer wavelength, resulting in a higher effective diffraction efficiency, and the buildup of noise owing to photovoltaic effects and photoinduced scattering is greatly reduced compared with single-photon recording in Fe-doped $LiNbO_3$ crystals using green light. Such improvement lowers the bit-error rate (BER) or, equivalently, increases system capacity. Furthermore, recent progress in compact, tunable diode laser sources has made recording in the near-infrared more attractive, as it allows for compact and inexpensive systems. In general the two-color technique is an elegant approach to solving difficult material issues. In digital holographic storage [17] information is typically encoded in the form of data pages, which includes modulation coding, data interleaving and error correction. The selective properties of volumetric gratings allow for a multitude of holograms to be recorded in virtually the same volume through angular, wavelength, or shift and correlation multiplexing techniques. Holographic digital data storage has the potential for extremely high data transfer rates (due to parallel organization of the information in the form of data pages) as well as high data densities due to the volumetric, 3-dimensional nature of the recording process.

The digital system (described in detail in [7]) consists of a 1 W Ti:sapphire laser pumped by a multi-line argon laser and tuned to 800 nm, a chrome-on-

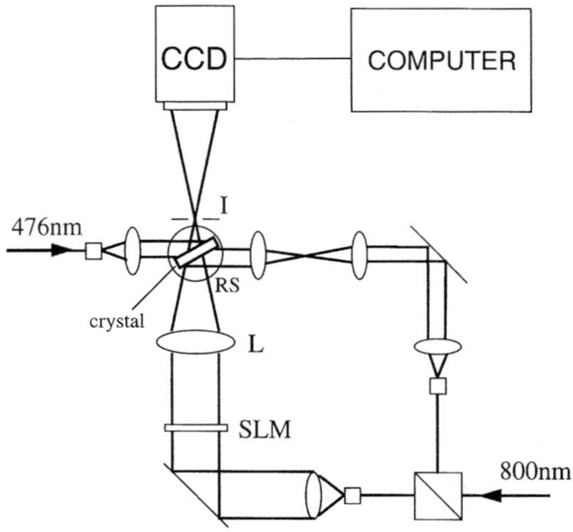

Fig. 10. Schematic of the two-photon digital holographic storage system: I, iris; L, imaging lens; RS, rotation stage; SLM, spatial light modulator

glass transmission mask, and a Kodak XHF CCD camera (Fig. 10). Gating light from another argon laser at 476 nm is delivered to the crystal by an optical fiber. The mask pattern contained 512 × 512, 18-μm square random data bits (100% fill factor) and alignment cross-hairs but did not include any modulation or error-correction codes. An off-the-shelf multi-element lens optimized for one-to-one imaging mapped the mask pixels onto the CCD pixels. The 18- μm SLM pixels are centered on single 9-μm square CCD pixels to reduce interpixel cross talk and relax misalignment tolerances. The surrounding nonactive CCD pixels act as a guard area that is equivalent to a smaller fill factor, and no oversampling is used for postprocessing. The cross-hairs are used during the manual alignment for accurate adjustment of tilt and magnification. Closed-loop computer control of the CCD camera x-y position provides fine alignment ($\approx 1\,\mu$m) of the cross-hairs before readout. The data are downloaded from the CCD to the controlling computer and adjusted by a local threshold algorithm: Each pixel value is normalized with respect to the intensity of 0's and 1's of its nearest crosshair to compensate for intensity variation across the page. The resulting histogram is used to evaluate the BER. We fitted the tails of the distributions of 0's and 1's to Gaussians to evaluate their overlap and the corresponding BER for the optimal threshold. A two-photon sensitive 1-mm-thick partially reduced $LiNbO_3$ crystal of near-stoichiometric ($[Li_2O] = 49.65$ mol%) composition [5,6] (doped with non-photoactive 0.2% Pr) was used in these experiments. The digital holograms of the mask were recorded at different wavelengths in the transmission geometry by use of e-polarized recording beams and 100 mW/ cm^2 of grating light (Fig. 11). The imperfect parallelism of the cut crystal caused distortions of the order of one CCD pixel across the page, which were compensated for

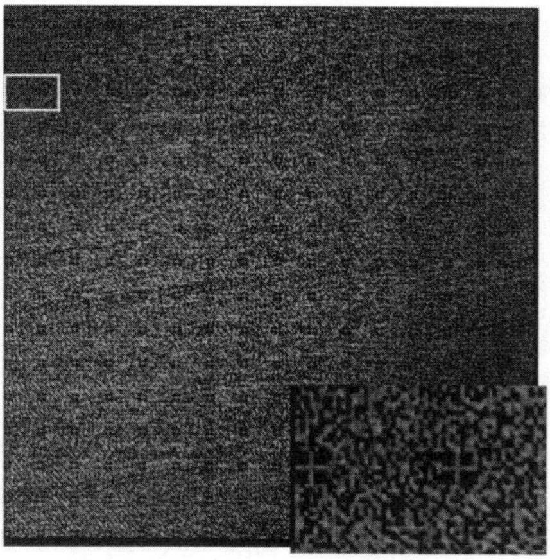

Fig. 11. Reconstructed hologram of a 256-kbit random data page and a blowup of a small region (*marked in white* on the page). The white cross-hairs are used for alignment and local thresholding

by automated realignment of the camera on the cross-hairs while the hologram was scanned upon readout. *A priori* knowledge of the fixed distortion can also be implemented in the signal-processing algorithm, at the expense of greater processing complexity and slower readout rates. In practical systems the problem can be eliminated with better crystal polishing or using distortion-compensating schemes, such as phase conjugate readout [18,19].

More than 1% diffraction efficiency was achieved for one page written to saturation. No attempt was made to optimize the position of the Fourier plane or the intensity ratio between the writing beams. We anticipate higher diffraction efficiency with further optimization, including the use of phase masks to record in the Fourier plane. The diffracted light signal shows no decrease in strength after several hours of continuous reading, and there is no degradation in the BER. The residual erasure in the near-infrared at writing intensities was two orders of magnitude slower than in the presence of the gating light. No dark decay was observed over several days. For the 256-kbit hologram shown in Fig. 11, the estimated BER is found to be 8×10^{-5} (Fig. 12), which is sufficient to guarantee a user BER of 10^{-12} when established modulation and error-correction codes [14,16] are implemented.

High scattering noise (equivalent diffraction intensity $\eta \approx 10^{-5}$) was the limiting factor in the number of holograms that could be multiplexed with reasonable BER. This noise floor could be significantly lowered by the use of antireflection coating and improvements in the growth and optical quality of stoichiometric $LiNbO_3$ samples. The preliminary results described here illustrate the high-resolution recording and high-capacity potential of a nonvolatile all-optical holographic data storage system.

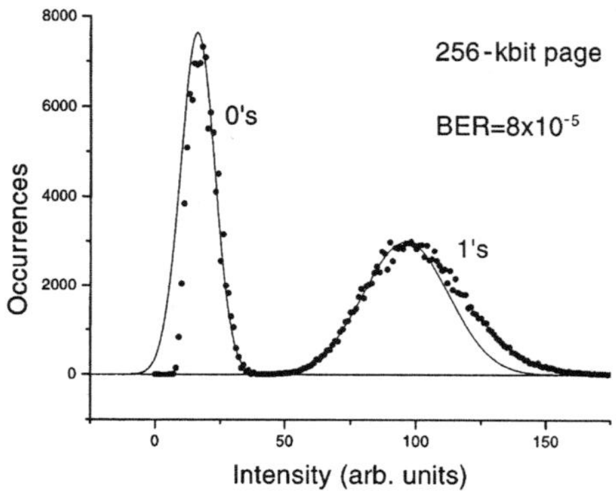

Fig. 12. Intensity distribution of data bits contained in a 256-kbit hologram after the intensities are normalized to neighboring cross-hairs. The high-intensity tail of the 0's and the low-intensity tail of the 1's are fitted to Gaussian functions. Their overlap is then calculated to yield a best-threshold BER estimate

6 Conclusions

Substantial improvement in two-photon recording sensitivity can be achieved in stoichiometric partially reduced lithium niobate. Two-photon sensitivity exhibits threshold-like behavior with respect to host crystal composition. The two-photon sensitivity and dynamic range of lightly reduced near-stoichiometric (for $[Li_2O] \gtrapprox 49.6\,mol\%$ in the crystal) lithium niobate is almost three orders of magnitude higher than that in as-grown congruent material. In undoped material small polarons are metastable shallow species responsible for the near-IR sensitivity, while bound bipolarons play the role of deep donors. In undoped material reduction induces bipolarons which provide efficient absorption for gating light and it also substantially increases the polaron lifetimes (to $\approx 1\,s$ and more) leading to greatly increased sensitivity. By introducing deep level dopants, such as Fe^{2+} and Mn^{2+}, in reduced crystals, the gating efficiency and readout erasure time can be increased further compared to undoped material. The major role of reduction is in the control of the density of the deep acceptor levels in doped material. Further increase in sensitivity and recording dynamic range can be envisaged in stoichiometric crystals codoped with both extrinsic deep donors and extrinsic shallow acceptor levels. Choices of potential dopant pairs may include, e.g., Mn^{2+}/Ce^{4+}, Mn^{2+}/Fe^{3+} [3], and Cu^{+}/Ce^{4+}. The preliminary results on recording of high-resolution digital holograms demonstrate good potential of nonvolatile gated two-photon materials for high-capacity all-optical holographic data storage systems.

References

1. D. von der Linde, A. M. Glass, K. F. Rodgers, Appl. Phys. Lett. **25**, 155 (1974)
2. Y.-S. Bai and R. Kachru, Phys. Rev. Lett. **78**, 2944 (1997)
3. D. L. Staebler, W. Phillips, Appl. Phys. Lett. **24**, 268 (1974)
4. K. Buse, A. Adibi, D. Psaltis, Nature **393**, 665 (1998)
5. S. S. Orlov, A. Akella, L. Hesselink, R. R. Neurgaonkar: High sensitivity two-photon non-volatile recording in lithium niobate, *Conf. Lasers Electro-Optics 1997*, Baltimore, MD, (Optical Society of America) postdeadline paper CPD29
6. L. Hesselink, S. S. Orlov, A. Liu, A. Akella, D. Lande, R. R. Neurgaonkar, Science **282**, 1089 (1998)
7. D. Lande, S. S. Orlov, A. Akella, L. Hesselink, R. R. Neurgaonkar, Opt. Lett. **22**, 1722 (1997)
8. O. F. Schirmer, O. Thiemann, M. Wöhlecke, J. Phys. Chem. Solids **52**, 185 (1991)
9. P. Nagels, in C. L. Chien and C. R. Westgate (Eds.): *The Hall Effect and its Applications* (Plenum, New York 1980)
10. J. Koppitz, A. I. Kuznetsov, O. F. Schirmer, M. Wöhlecke, B. C. Grabmaier, Ferroelectrics **92**, 233 (1989)
11. A. Mehta, E. K. Chang, D. M. Smyth, J. Mater. Res. **6**, 851 (1991)
12. Stoichiometric lithium niobate crystals with varying Fe concentrations grown by continuous charge double crucible Czochralski (DCCZ) method were provided to us by the National Institute for Research in Inorganic Materials in Japan (Dr. K. Kitamura)
13. F. Jermann, M. Simon, E. Krätzig, J. Opt. Soc. Am. B. **12**, 2066 (1995)
14. J. R. Carruthers, G. E. Peterson, M. Grasso, P. M. Bridenbaugh, J. Appl. Phys. **42**, 1846 (1971)
15. L. Kovacs, M. Wöhlecke, A. Javanovic, K. Polgçr, S. Kapphan, J. Phys. Chem. Solids **52**, 979 (1991)
16. P. F. Bordui, R. G. Norwood, D. H. Jundt, M. M. Fejer, J. Appl. Phys. **71**, 875 (1992)
17. J. F. Heanue, M. C. Bashaw, L. Hesselink, Science **265**, 749 (1994)
18. L. Hesselink, S. Redfield, Opt. Lett. **13**, 877 (1988)
19. G. W. Burr, J. Ashley, H. Coufal, R. K. Grygier, J. A. Hoffnagle, C. M. Jefferson, B. Marcus, Opt. Lett. **22**, 639 (1997)

Photorefractive Copper-Doped $LiNbO_3$ Waveguides for Holography Fabricated by a Combined Technique of Ion Exchange and Ion Implantation

Sergey M. Kostritskii[1] and Paul Moretti[2]

[1] Physics Departement, Kemerovo State University
Krasnaya str. 6, Kemerovo, 650043 Russia
skostritskii@yahoo.com

[2] Laboratoire de Physico-Chimie des Matériaux Luminescents
Université Claude Bernard Lyon I, 69622 Villeurbanne, France
pmoretti@pcml.univ-lyon1.fr

Abstract. We report on the fabrication and properties of photorefractive waveguides in pure or magnesium-doped lithium niobate. He^+ implantation is used to form planar waveguides while an ion exchange process performs copper doping of the substrates surface. The photorefractive response of waveguides fabricated by using the two possible sequences, first ion implantation and second copper doping, and the reverse, are compared. High values of the light-induced refractive-index change and holographic sensibility are obtained. We show that beam fanning is dramatically suppressed by heavy magnesium co-doping. Fanning-free propagation is therefore possible for a wide range of in-coupled power. Steady-state diffraction efficiency of 30% is demonstrated, but higher values could be reached by proper choice of interaction length in the waveguides.

1 Introduction

Among the promising inorganic nonlinear crystals competitive to semiconductors for optical applications, lithium niobate ($LiNbO_3$) is presently the most widely used substrate because of its outstanding acousto- and electro-optical properties and its large commercial availability. Furthermore, with appropriate doping, i.e. essentially by iron or copper, very important induced photorefractive [1] effects can be also used, not only in bulk devices, as for holographic volume storage, and especially for communication signal processing [2], but also in integrated optics, like optical switching, holographic resonant filtering, DBR (Distributed Bragg Reflector) or memory [3]. The formation of highly efficient holographic gratings has been thus demonstrated in lithium niobate (LN) waveguides [4,5,6]. The fabrication of photorefractive waveguides (PRW) is of great interest since it provides compatibility with other miniaturized integrated-optical devices which are in common use today, in particular, laser diodes and optical fibres. Additionally, the optical confinement inherent to waveguide structures allows to maintain high

P. Boffi, P. Piccinin, M. C. Ubaldi (Eds.): Infrared Holography for Optical Communications, Topics Appl. Phys. **86**, 59–74 (2003)

intensity within the waveguide, and the resulting increased intensity-length product leads (for a given input power) to an appreciable decrease in the effective photorefractive response time [7]. Moreover, when holographic components are recorded by guided beams, wavefront distortions introduced by the mode structure or imperfections of the waveguide can automatically be compensated [8]. Thus, Bragg reflectors as small as the waveguide dimensions, so-called "micro-Bragg reflectors", can be fabricated in PRW allowing us to design small complex optical systems for massively parallel signal interconnection for the next generation of signal processors [9]. However, manipulating photorefractivity in lithium niobate by guided beams generally involves the use of high light intensity during holographic recording [10], which unfortunately provides good conditions for beam fanning effects to appear [11,12]. Such an extra nonlinear effect may play an undesirable competing role. It was established in fact to be a key factor degrading the signal-to-noise ratio of the photorefractive holographic storage in bulk crystals [11] and even suppressing holographic recording in planar optical waveguides [12]. Obtaining efficient PRWs in $LiNbO_3$ faces to two main problems: on the one hand, achieving the correct doping, in order to achieve a large concentration of sensitive impurities such as iron or copper, and on the other hand to elaborate "good" waveguides, i.e. by using methods which retain the fundamental properties of the bulk, and particularly the electro-optical ones. Surface doping methods, like in-diffusion or ion-exchange, are undoubtedly preferable to melt doping during crystal growth, because a significant increase of the impurity content has been demonstrated in both cases [11,12,13,14,15,16]. These methods are more sophisticated, but more promising. They are very accommodating since they can be directly combined with the related waveguide fabrication process [6,14,15]. One advantage of the exchange process must, however, be outlined. It is as a low-temperature process ($< 300\,°C$), and the new method of combined proton and copper exchange recently reported [15,17] is particularly attractive because it can also be applied to low-curie-temperature compounds like $LiTaO_3$ [18]. However, it has been shown that the electro-optic coefficients (EOCs) are degraded in proton-exchange waveguides [17,19], now it is well known that large EOCs are of great advantage for holographic recording. In contrast, no degradation of the EOCs is observed in waveguides fabricated by light ion implantation [20], and furthermore, high efficiency photorefractive waveguides have been obtained in SBN [21] or $BaTiO_3$ [22]. Therefore, to improve the photorefractive properties of copper-doped $LiNbO_3$ waveguides; we recently proposed a new two-step method combining the two techniques of ion implantation and ion exchange [23]. Our purpose is (i) to form single-mode optical waveguides by ion implantation and (ii) to increase the photorefractive sensitivity of the crystal by performing a combined proton and copper exchange process in specific conditions, namely by minimizing the proton exchange in order to avoid any significant degradation of the electro-optical effect in the crystal. Note that only weak changes of the effective refractive index, insufficient to support any guided mode, are expected due to the exchange process itself. There are actually two possible ways to proceed: one is to implant first and subsequently to dope the surface by the exchange process, i. e. (i) then (ii) (Meth I), the other is to reverse these steps, i.e. (ii) then (i) (Meth II). In this paper we compare the photorefractive performance of waveguides fabricated by both methods (Meth I and Meth II). The light-induced refractive-index change in saturation state, the beam-fanning effect and the response time are investigated with respect to their dependence on the exchange process parameters in nominally pure and magnesium-doped $LiNbO_3$ substrates. High diffraction efficiency is demonstrated in Mg co-doped crystals, suggesting developing our new methods for fabricating micro-Bragg reflectors in $LiNbO_3$ waveguides.

2 Waveguide Fabrication and Characterization

In contrast to the proton exchange or diffusion methods [24], which can be used to form waveguides in very few crystals only, light ion implantation (He^+ or H^+) can be applied to fabricate waveguides efficiently in numerous crystals [24,25,26,27,28,29,30], including $LiNbO_3$ [20,28,29] and other photorefractive waveguides [21,22,30]. Note that ion implantation can be performed at room temperature or low temperature. The principle of the method can be summarized in the following way. The damaging effect due to the nuclear collision processes between the incoming ions (He^+ or H^+) and the target atoms generate an optical barrier in crystals (negative local index change) whose thickness and position beneath the surface depend on the ion energy. We use successively three helium implantations (at room temperature), at slightly different doses and energies, to tailor approximately rectangular-like damage profiles. The resulting ion concentration profiles given in Fig. 1, calculated by Transport Ion in Master Code (TRIM) or Profile-Code simulation [31], is very closed to the damage profile and is a good indication of the resulting optical barrier shape. The energy and doses used are indicated in Fig. 1. Multiple implantation enables us to enlarge the thickness (here from 0.3 to 0.7 μm) of the optical barrier, which results in a better confinement of the light in the waveguides.

Nominally pure (LN) or Mg-doped (Mg-LN) optical grade $LiNbO_3$ substrates (Y- and Z- cuts), typically of 5 mm × 10 mm × 1 mm dimensions are used. After implantation the waveguides are weakly annealed (200–230 °C, 30 min) to decrease the propagation losses. Measurements of the effective indices of the implanted waveguides by dark-line spectroscopy at 632.8 nm gives extraordinary index values typically in the range $2.210 \leq n_e \leq 2.215$. For extraordinarily polarization, i.e. TM and TE mode, in the Z-cut and in the Y-cut samples, respectively, the waveguides are monomode for the excitation wavelengths of 441.6 and 514.5 nm, confirmed by bright line spectroscopy. In contrast, the waveguides are multimode (supporting two or three modes,

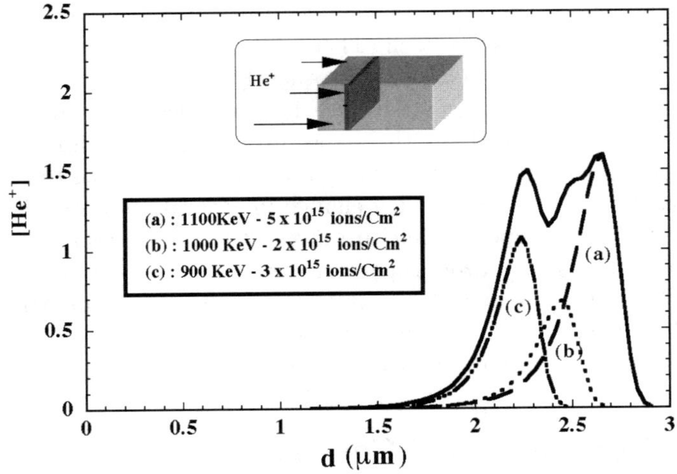

Fig. 1. Ion implantation. He^{+} ions concentration profiles ($[He^{+}]$,10^{20} At/cm^3) calculated for multiple implantation in $LiNbO_3$ with a total dose of 1×10^{16} ions/cm^2 (Profile-Code simulation program). The plain curve is the summation of (**a**), (**b**) and (**c**). The principle of waveguide formation is shown schematically in the insert

including leaky modes) for ordinary polarization. There is no fundamental difference regarding the waveguiding properties between the Y- and Z-cut samples.

3 Copper Doping by the Exchange Process

In contrast to previous work, where a waveguiding proton-exchanged layer where formed and subsequently doped [14,15,16,17], in this study a combined proton and copper exchange is performed in a single melt of benzoic acid mixed with lithium benzoate (LB) and copper acetate. It is well known that the presence of LB in a molten benzoic acid acts as a buffer in the lithium–proton exchange process, and at the content level of LB that we use some proton exchange can still be observed, but not enough to induce a meaningful change of the extraordinary refractive index. The copper and hydrogen content of the samples are determined by optical absorption spectroscopy and on the basis of previous data obtained on combined proton- and copper-exchanged waveguides [23]. The exchange-process parameters are varied to determine the optimal conditions that produce the largest photorefractive response and/or holographic recording. However, the conditions of the exchange process differ according to the fabrication sequences of the photorefractive waveguide, i.e. Meth I or Meth II. In Meth I the combined proton and copper exchange is performed by treatment in molten benzoic acid mixed with LB and copper acetate at concentration levels varying from 3.0 to 3.6 mol% and 0.8 to 4 mol%, respectively, after the preparation of the waveguides by ion implantation. The duration of the treatments ranged from 20 to 40 min at 230 °C. This low temperature process is necessary, in order to be sure not to destroy the optical barrier formed by ion implantation [25]. The structure of the exchanged layer is almost entirely composed of the α phase of the $Li_{1-x}H_xNbO_3$ system, with low x values ranging approximately from 0.005 to 0.03, and the copper concentration is in the range of 0.012–0.06 mol%. In Meth II the implanted waveguides are fabricated subsequently to the exchange process. There is therefore no need for any temperature restriction. The substrates are dipped in molten benzoic acid mixed with 0.1–0.15 mol% Cu2O and 1 mol% lithium benzoate, at 230–249 °C, for 60–90 min, then a strong annealing is applied in dry air at 390 °C for 10 – 16 h. Consequently, the proton concentration is insignificant while high copper concentration can be achiebed, from 0.08 to 0.5 mol%.

4 Photorefractive Properties

The photorefractive properties of the various waveguides were investigated by waveguide Raman spectroscopy to determine their dependence on the fabrication method, i.e. the sequential order of the processes, the copper and proton exchange conditions and the effect of co-doping the $LiNbO_3$ substrates with magnesium. The aim was to optimize the waveguides for holographic recording.

4.1 Method of Characterization

Waveguide Raman spectroscopy is used as a characterization technique. The great effectiveness of this method has been demonstrated [15]. The technique is unique in providing a 1 μm spatial resolution both laterally and in the direction perpendicular to the guided surface. As described schematically in Fig. 2, a focused laser beam is coupled into the waveguides with aid of a rutile prism. Depending on the angle of the light entering through the prism, different extraordinarily polarized guided modes (TM) are excited. Optimal conditions of excitation are achieved by precision adjustment of a selected guided mode with aid of modal bright-line spectroscopy performed at a low input power. The Raman intensity (I) is measured at the right angle to the waveguide surface with the aid of a "Ramanor U1000" spectrometer. Raman scattering, where selection rules depend on the direction and polarization exciting light [32] can be indeed efficiently used as a probe of the fanning and polarization state of the propagating guided beam.

In absence of photorefractive activity I is classically independent of the excitation time. However, according to previous work [15], when the photorefractive effect occurs, the Raman intensity I of any phonon line shows a time exponential decay with a time constant τ , due to the spreading of the laser beam into a fan. The observed time dependence of I coincides with the kinetics of the build-up of the photorefractive effect and τ depends on the copper

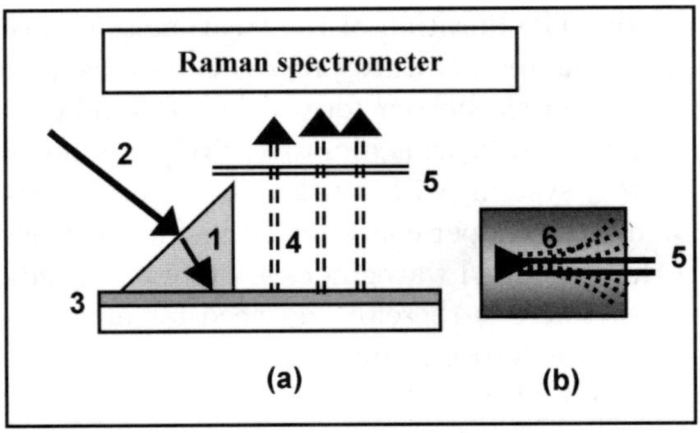

Fig. 2. Principle of photorefractive properties investigation by waveguiding Raman spectroscopy: (**a**) prism coupling (1) of a focused laser beam (2) in the waveguide (3) and detection (at 90°) of the Raman scattering (4) through a slit (5), (**b**) top view, beam-fanning effect (6)

concentration in the waveguide. Because of the beam fanning the Raman intensity I decreases, therefore, down to a steady-state value which is proportional to the light-induced extraordinary refractive-index change (Δn_s). The relative change of the Raman signal amplitude, $I_0 - I_s/I_0$, where I_0 is the Raman intensity at the initial stage (at $t = 0$) and I_s is the intensity in steady state (at $t \to \infty$), depends directly on Δn_s, as follows [15]:

$$I_0 - \frac{I_s}{I_0} = A\frac{\Delta n_s}{1 + A\Delta n_s}, \tag{1}$$

where A is a coefficient related to the conditions of Raman measurement, which is here a constant since all the experimental conditions are identical.

4.2 Effects of the Combined Copper and Proton Exchange Conditions

From the Raman spectroscopy measurements, two main photorefractive parameters are determined in the waveguides: the magnitude of the saturated index change (Δn_s) and the kinetics of the photorefractive response, i.e. the time constant τ. The latter enables us to calculate the photorefractive sensitivity S defined by the following relation:

$$S = \frac{\Delta n_s}{\tau J}, \tag{2}$$

where J is the density of light in the waveguide. As a characteristic example of the role of the exchange conditions, the dependence of Δn_s on the copper concentration in waveguides, fabricated by Meth I in LN substrates is given in Fig. 3 for two levels of residual hydrogen concentration (x). The Δn_s values are deduced from Raman measurements at $\lambda = 441.6$ nm. The differential optical density δD_{Cu} is the optical absorption density measured (at the

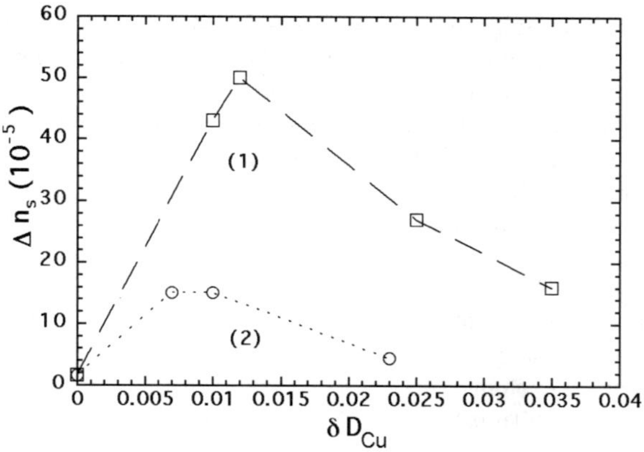

Fig. 3. Light-induced refractive-index change (Δn_s) in saturation state versus the copper-induced increment of the optical density (δD_{Cu}) measured by absorption spectroscopy; δD_{Cu} is proportional to the copper content in the samples. The *lines* are merely a guide for the eye. These *lines* correspond to two different levels of hydrogen concentration, x_I (1) and x_{II} (2) with $x_I < x_{II}$ in the waveguiding layer after the combined proton and copper exchange (Meth I, with pure substrates)

excitation Raman wavelength) in the UV–visible absorption spectra of the samples, by using a non-exchange waveguide as reference, and is therefore proportional to the copper concentration.

Figure 3 shows two striking features: (1) the variation of Δn_s with the copper concentration reaches a maximum for intermediate values of δD_{Cu}, (2) there is a crucial influence of the hydrogen concentration x on the Δn_s magnitude, i.e. for a given Cu concentration, Δn_s is dramatically reduced for the highest proton concentration. The first behaviour can be explained by a quenching effect of the photorefractive activity in the waveguide due to the increase, beyond a certain threshold, of the optical absorption with the copper concentration [23]. The second one must be attributed to the degradation of the electro-optical coefficient r_{33} when x increases [16,17].

The effects of the parameters of the combined exchange process on the photorefractive properties for some waveguides fabricated by Meth I are illustrated in Table 1. The S and Δn_s values obtained for a given launched intensity ($J_{in} = 12 \times 10^4\,W/m^2$) are summarized, together with the optical absorption data and the exchange-process conditions. The results show clearly that the photoinduced refractive-index change Δn_s is very dependent on the exchange conditions and that the copper doping increases dramatically the photorefractive sensitivity S of the He-implanted waveguides. A factor of the order 1000 is obtained here, but note that by decreasing the thickness of the implanted waveguides higher factors (up to 3000) can be reached [23].

Table 1. Results obtained for He-implanted waveguides subsequently doped by a combined proton and copper exchange process. The total dose (multiple implantation is performed, see Fig. 1) is 1.0×10^{16} ions/cm^2 and 2.0×10^{16} ions/cm^2 for the c, d samples and the others, respectively. Reported are: the cut direction, the exchange parameters, the optical density δD_{Cu} of the copper-induced band at 380 nm, the saturated value of the light-induced refractive-index change Δn_{s} (in 10^{-5}) and the estimated values of the photorefractive sensitivity S

Waveguide	Cut	Concentration of lithium benzoate (mol%)	Concentration of copper acetate (mol%)	Duration of ionic exchange (min)	δD_{Cu}	$\Delta n_{\mathrm{s}}/10^{-5}$	S/m^{2}J^{-1}
a	*Y*	0	0	0	0.000	1.6	3×10^{-12}
b	*Z*	3.4	0.8	30	0.007	15	7.9×10^{-10}
c	*Y*	3.0	2.0	40	0.023	4.4	6.7×10^{-11}
d	*Z*	3.4	4.0	30	0.025	27	8.6×10^{-10}
e	*Y*	3.4	2.0	20	0.012	50	2.4×10^{-9}

4.3 Effects of the Fabrication Method and Mg Co-doping

As described in Sect. 4.1, the beam fanning decreases the Raman intensity I by an amount which is proportional to the light-induced extraordinary refractive-index change Δn_{s} in the steady state. Since Δn_{s} increases with the input power intensity J, the dependence of I on J has a pronounced sub-linear character when Δn_{s} reaches appropriate high values. To characterize the beam fanning in the various waveguides fabricated by both methods, in LN and Mg-LN, the dependence of the steady-state output Raman intensity I (at $\lambda = 514.5\,\mathrm{nm}$) on the in-coupled laser power P_{in} is measured for fixed experimental conditions. The results are shown in Fig. 4. The stronger is the divergence of this dependence from a linear law, the larger is the photorefractive effect. Clearly, the photorefractive properties depend dramatically on both the fabrication method and the nature of the $LiNbO_3$ substrate. The beam fanning of the guided light is the largest one in the waveguides fabricated by Meth II and in LN. Larger Δn_{s} and faster response time τ are also observed in the Meth II waveguides, demonstrating larger photorefractive sensitivity. The dependence of Δn_{s} on the fabrication method and Mg co-doping is shown in Table 2. The highest measured Δn_{s} values are observed in waveguides fabricated in LN substrates, in agreement with the known quenching effect of Mg on the photorefractive properties of $LiNbO_3$.

It is worth noting that fanning only appears beyond a certain threshold power ($P_{\mathrm{in},t}$). The $P_{\mathrm{in},t}$ value is very low ($P_{\mathrm{in},t} < 5 \times 10^{-5}\,\mathrm{W}$) and inversely

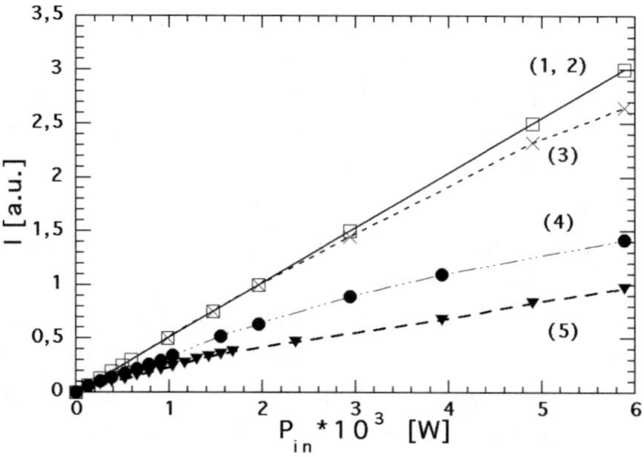

Fig. 4. Steady-state Raman scattering intensity (I) for line of the E(TO,1) mode versus in-coupled laser power (P_{in}) measured in the different He-implanted waveguides: (1) undoped waveguides fabricated in nominally pure $LiNbO_3$ substrates; (2), (3) Cu-doped waveguides in $LiNbO_3$-Mg substrates fabricated by Meth I (2) and Meth II (3); (4), (5) Cu-doped waveguides in nominally pure $LiNbO_3$ substrates fabricated by Meth I (4) and Meth II (5)

Table 2. Photorefractive properties of the waveguides with respect to the fabrication method and the nature of the substrate. The samples are denoted according to the labelling used in Fig. 4. Indicated are: the specification of substrates, the fabrication method, the threshold energy of fanning, the steady-state refractive-index change $\Delta\, n_u s$ evaluated from Raman data on beam fanning at 514.5 nm excitation, and the maximum value of steady-state diffraction efficiency η_s of phase holograms recorded at 632.8 nm. The η_s value of the 2, 4 samples coincides in fact with the accuracy of our measurements, but holographic recording is visually evident

Waveguide	$LiNbO_3$ substrate	Method	Threshold $P_{in,t}$/ mW	$\Delta\, n_s/10^{-5}$	$\eta_s/\%$
1	LN	-	≥ 5	< 2	0
2	Mg-LN	I	≈ 5	2	≤ 0.1
3	Mg-LN	II	1.4	13	30
4	LN	I	0.1	112	≤ 0.1
5	LN	II	< 0.05	203	0.6

proportional to the copper concentration in LN waveguides, and especially in those fabricated by Meth II. In contrast, waveguides in Mg-LN exhibit much a higher threshold $P_{\text{in},t}$ that varies from 10^{-3} to 6×10^{-3} W (in dependence on the Cu-doping conditions) and the effect of the method is less pronounced. Thus, fanning-free propagation of light can be performed for a wide range of coupled power in the waveguides implanted in Mg-LN. These results are in agreement with the marked progress recently claimed in optimizing Fe-doped $LiNbO_3$ bulk crystals for three-dimensional photorefractive storage achieved by co-doping the crystals with optical-damage-resistant impurities (Mg, Zn, In) [11]. Moreover, when strong fanning occurs, we observe that the output Raman intensity has a chaotic fluctuating behaviour, indicating clearly that photorefractive parametric amplification of noise light is the essential mechanism involved in the observed fanning effect. Note that such a mechanism has been established to be specific for the light-induced scattering in proton-exchanged $LiNbO_3$ waveguides at high in-coupled intensity [34]. Finally, according to the Δn_s values reported in Table 2, highly efficient holographic recording is expected, especially in He-implanted waveguides fabricated by Meth II in non-Mg-doped substrates, since it is usually believed that the larger is the photorefractive effect, the stronger is the two-beam coupling. However, as demonstrated below, the fanning may produce an undesirable competing nonlinear effect.

5 Holographic Recording

Holographic gratings are written and read by in- and out-coupling two slightly focused laser beams ($\lambda = 632.8$ nm) through a $Bi_{12}GeO_{20}$:Al prism. In these experiments two modes intersect at an angle of $2\Theta = 10°$, and the interaction length in the waveguides is about 0.2 mm. Such a low value is chosen in order to avoid the undesirable influence of TM-to-TE mode conversion in our holographic recording experiments with high-power guided beams [32,33]. Furthermore, the interaction area is adjusted to be directly under a coupling point of the prism-air-waveguide interface, which is in principle not sufficient to give rise to any marked conversion. During the build-up of the refractive-index grating, the diffraction efficiency η is measured as a function of time by blocking one of the two in-coupled beams for a short time and measuring the ratio of the diffracted and total light intensity of the out-coupled beams.

5.1 Dynamics of Holographic Recording

In He-implanted waveguides fabricated in both LN and Mg-LN substrates, whatever the in-coupled power level for the red line of the He-Ne laser we use, we do not observe holographic grating recording. However, in the Cu-doped waveguides, effective photorefractive recording of phase holograms becomes possible. The combined techniques proposed here, especially Meth II, allows the fabrication of waveguides showing a sufficiently high level of diffraction efficiency η for recording phase holograms. The temporal evolution of η during holographic recording shows two types of characteristic behaviour, depending on the in-coupled power and the nature of the waveguide, i.e. in fact on the beam-fanning conditions. They are illustrated in Figs. 5a,b for waveguides fabricated by Meth II in LN and Mg-LN, respectively. A surprising result is that for strong fanning ($P_{\text{in}} > P_{\text{in},t}$, see Sect. 4.3) instead of an exponential-like temporal law, a rather fast rise is observed followed by a slower decay towards a steady-state value η_{s}. Since it is well established that the fanning strength depends on the exposure time, this behaviour can actually be easily explained. In the initial stage, while fanning is not significant, the effective holographic recording can occur, but increasing the exposure time enhances the fanning strength and prevents subsequent holographic recording. Furthermore, the fanning induces photoerasure of holograms recorded in the initial stage because of the trajectory change of the light in the planar waveguide.

For a given waveguide, the occurrence of a transient diffraction efficiency peak occurs only for a certain range of in-coupled power (ΔP_{in}). Below this range, the diffraction efficiency follows the usual exponential-like law; above ΔP_{in}, no holographic recording is observed, because it is entirely suppressed by the extremely strong fanning effect. This power range greatly depends on the nature of the waveguide: ΔP_{in} is very large in Mg-LN, from 3×10^{-3} W up to a value larger than the highest we could use in our experiments, while ΔP_{in} only extends from 0.2×10^{-3} to 1.5×10^{-3} W in LN substrates. The optimal value of the in-coupled power P_{in} to achieve the most efficient holographic recording in a given waveguide could be defined as the low-threshold of ΔP_{in}.

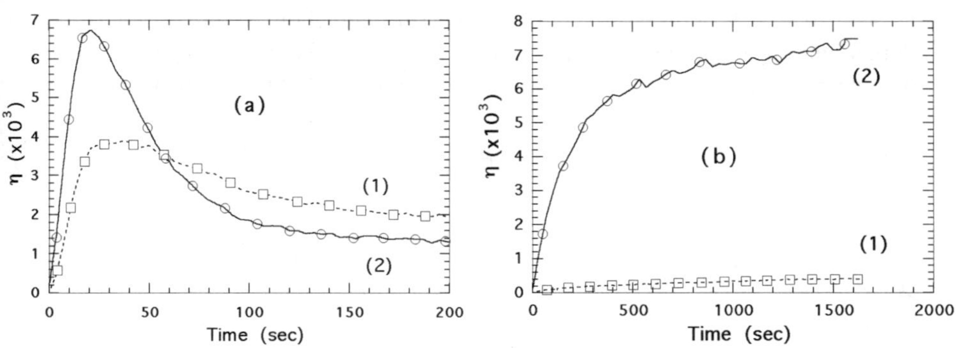

Fig. 5. Kinetics of diffraction efficiency (η) during build-up of refractive-index grating (**a**) strong fanning case ($P_{\text{in}} > P_{\text{in,t}}$, see text) for a waveguide fabricated by Meth II in a pure $LiNbO_3$ substrate, the laser wavelength is 632.8 nm, the in-coupled power Pin is 6×10^{-5} W (1) and 10^{-4} W (2) (**b**) lacking fanning case ($P_{\text{in}} < P_{\text{in,t}}$, see text) for a waveguide elaborated by Mth II in an Mg co-doped $LiNbO_3$ substrate, the Laser wavelength is 632.8 nm, the in-coupled power Pin is 5×10^{-4} W (1) and 1.8×10^{-3} W (2)

It should be noted, that the significant difference of the optimal in-coupled power observed between the Mg-doped and undoped samples is in contrast to previous data [11] reported on the influence of damage-resistant dopants on the holographic recording process in bulk Fe:$LiNbO_3$. Such a discrepancy may be attributed to the specific intensity-dependence of the photorefractive effect observed in copper-doped and proton-exchanged $LiNbO_3$ waveguides [23]. Our results give further proof of the occurrence of a threshold effect with respect to the incident light intensity [11] for photorefractive light-induced scattering and of the existence of a certain optimum pump light intensity to achieve holographic recording with maximum efficiency. In our He-implanted waveguides this efficiency should be easily controlled over a wide power range by Cu and Mg doping at suitable concentration levels.

5.2 Diffraction Efficiency

The steady-state value is found to be crucially dependent on the conditions of the proton-assisted copper exchange, post-exchange annealing, and even more on Mg co-doping. Table 2 provides a comparison of the measured η_s values. It must be pointed out that: (1) η_s is larger in waveguides fabricated by Meth II, (2) the largest value, $\eta_s = 30\%$ (sample 3 in Table 2), is obtained in substrates highly doped with Mg ($C_{Mg} = 7\,mol\%$) and processed by the most intensive copper exchange and the longest annealing time at 390 °C. Note that this diffraction efficiency value is, to our knowledge, the highest reported for holographic recording with red light in a PRW fabricated in a $LiNbO_3$ crystal.

Another striking feature is the non-monotonous dependence of η_s on P_{in}, as seen in Fig. 6. Therefore, there is a certain threshold value P_s where the dependence $\eta_s(P_{in})$ changes qualitatively, i.e. for $P_{in} > P_s$ the value of η_s decreases with larger in-coupled power P_{in}. This behaviour can be assigned

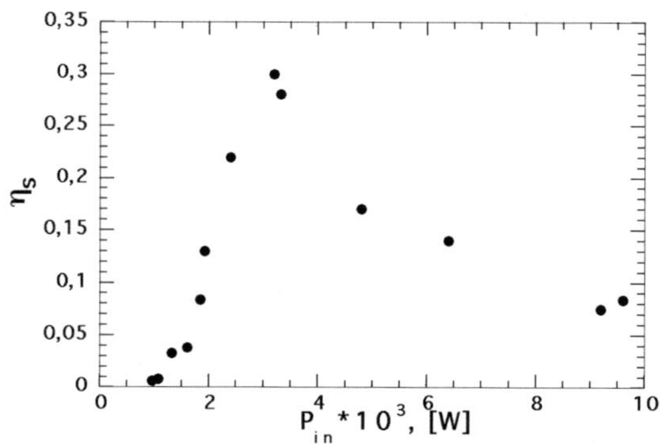

Fig. 6. Steady-state diffraction efficiency (η_s) versus in-coupled power (P_{in}) in a copper-doped waveguide fabricated by Meth II in $LiNbO_3$-Mg substrate (Sample (3) in Table 2 and Fig. 4)

to the influence of the fanning on the guided light. It has been observed both in bulk crystals [11] and in optical waveguides [12] and established a specific feature of fanning influence on holographic recording. The values of $P_{\mathrm{in},t}$ obtained in our Raman spectroscopy (see Sect. 4.3) and holographic experiments are different as a result of the specific spectral dependence of the photorefractive effect [23]. However, there is a clear correlation between the two values. The higher $P_{\mathrm{in},t}$ values in the Raman experiments corresponds to higher low-thresholds in the holographic measurements.

6 Discussion and Conclusions

Photorefractive waveguides have been fabricated in $LiNbO_3$ by a two-step method based on the combination of light ion implantation for waveguide forming and an ion-exchange process for copper doping. The influences of the exchange conditions, the order of the process steps, and Mg co-doping of the substrates on the photorefractive properties of the waveguides have been clarified. Compared to the simple technique of combined proton and copper exchange, the two-step method developed here shows a greater versatility, mainly because more fabrication parameters can be adjusted independently for optimizing the photorefractive properties of the waveguides. Varying the implantation parameters (dose and energy), for example, allows us advantageously to tailor, in some degree, the index profiles of the waveguides. The possibility of seperating the waveguide fabrication from the doping process permits better control of the electro-optical coefficients in the waveguides. Their change can be minimized by strongly reducing the residual proton concentration, as demonstrated in Sect. 4.2, and even could very likely be completely suppressed in Meth II waveguides, because a higher post-exchange annealing temperature can be used, that enhances the proton diffusion towards the bulk. This assumption is indirectly confirmed by the higher light-induced refractive-index change (Δn_{s}) reported in Table 2. The copper doping increases dramatically the photorefractive sensitivity of the He-implanted waveguides but a quenching effect of the photorefractive activity is observed beyond a certain Cu concentration (see Sect. 4.2). It is worth noting that Cu-doping does not induce any permanent shallow traps in the $LiNbO_3$ band gap [36]. This should be a great advantage in terms of the grating stability of our waveguides over Ti-indiffused Fe-doped $LiNbO_3$ ones, where fast dark compensation occurs [37] because of permanent shallow-filled traps involving Fe–Ti defect complexes [36]. Moreover, the higher the process temperature is, the better the in-depth copper diffusion in the crystal and the higher the Cu concentration in the waveguides. This is the second main advantage of processing Meth II. Consequently, better overlapping between the optical modes and the Cu concentration profile is probably obtained, resulting in a higher steady-state refractive-index change and diffraction efficiency, as seen in Fig. 4 and Table 2. Figure 7 illustrates schematically the

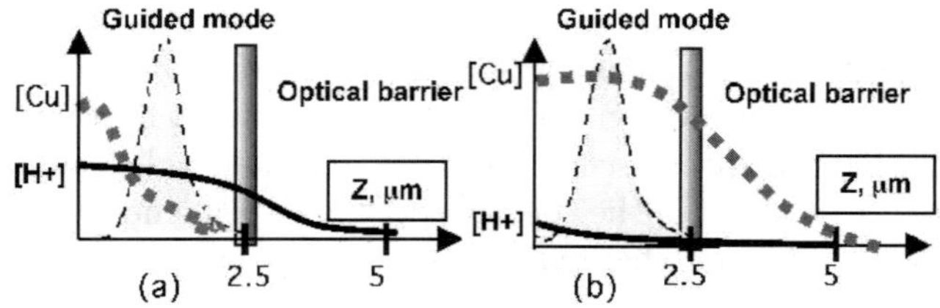

Fig. 7. Schematic diagram of the waveguide structures according to the methods used: (**a**) Meth I, fabrication of the waveguide by ion implantation subsequently doped by the proton-assisted exchange process; (**b**) Meth II, reversed process

differences between the two waveguide structures according to the fabrication methods.

The fastest response time τ of the beam-fanning effect is also observed in the Meth II-waveguides, demonstrating a larger photorefractive sensitivity factor S. With regard to the high Δn_s values attained in these waveguides fabricated in LN substrates, $\Delta n_s = 2 \times 10^{-3}$, the possibility of highly efficient holographic recording is expected. However, we do demonstrate that fanning of guided light is the main limiting factor for holographic recording of refractive-index gratings. We show that using $LiNbO_3$ substrates doped to a high Mg concentration level (up to 7%), although limiting the Δn_s change compared to pure substrates, allows us to overcome this problem and that beam-free propagation can be obtained over a wide range of in-coupled power (see Fig. 4). High diffraction efficiency is therefore demonstrated, $\eta_s = 30\%$, at He–Ne laser excitation wavelength. This high recorded value, with respect to the small interaction length of the guided beams used (0.2 mm) for recording the phase holograms, shows that micro-Bragg reflectors with controllable parameters should be easily fabricated by guided beams in waveguides fabricated by Meth II in Mg-LN substrates. Real future prospects do exist to extend the holographic performance of Mg co-doped waveguides in the near-infrared range for telecommunication applications, while the high holographic sensitivity demonstrated for small in-coupled power in waveguides fabricated in pure substrates is promising for the holographic amplification of weak optical signals.

References

1. A. Ashkin, G. D. Byd, J. M. Dziedzic, R. G. Smith, A. A. Ballman, J.J. Levinstein, K. Nassau, Appl. Phys. Lett. **9**, 72 (1966)
2. P. Boffi, M. C. Ubaldi, D. Piccinin, C. Frascolla, M. Martinelli, IEEE Photon. Technol. Lett., **12** 5 (2000)
3. E. Wood, P. J. Cressman, R. L. Holman, C. M. Veber: *Photorefractive Materials and their Applications II*, ed. by P. Günter, J.-P. Huignard (Springer, Berlin, Heidelberg 1989)

4. K. Buse, S. Reer, K. Peithman, S. Kapphan, M. Gao, E. Krätzig: Phys. Rev. B **56**, 1225 (1997)
5. J. Huckriede, D. Kip, E. Krätzig, Appl. Phys. B **68**, 1 (1999)
6. J. Huckriede, D. Kip, E. Krätzig, Appl. Phys. B **72**, 749 (2001)
7. K. E. Younden, S. W. James, R. W. Eason, P. J. Chandler, L. Zhang, P. D. Townsend, Opt. Lett. **17**, 1509 (1992)
8. L. Wan, Y. Yuan, G. Assanto, Opt. Commun. **74**, 361 (1990)
9. K. Itoh, K. Ikezawa, W. Watanabe, Y. Furuya, Y. Masuda, T. Toma, Opt. Express **2**, 503 (1998)
10. O. Matoba, K. Ikezawa, K. Itoh, Y. Ichioka, Opt. Rev. **2**, 438 (1995)
11. N. Y. Kamber, J. Xu, S. M. Mikha, G. Zhang, X. Zhang, G. Zhang, J. Appl. Phys. **87**, 2684 (2000)
12. S. M. Kostritskii, D. Kip, Phys. Stat. Sol. A **169**, 171 (1998)
13. K. Peithmannn, J. Huckriede, K. Buse, E. Krätzig, Phys. Rev. B **61**, 4615 (2000)
14. A. D. Novikov, S. G. Odoulov, V. M. Shandarov, S. M. Shandarov, Sov. Tech. Phys. **33**, 969 (1988)
15. S. M. Kostritskii, O. M. Kolesnikov, J. Opt. Soc. Am. B **11**, 1674 (1994)
16. D. Kip, F. Rickermann, E. Krätzig, Opt. Lett. **20**, 1139 (1995)
17. F. Rickermann, D. Kip, B. Gather, E. Krätzig, Phys. Stat. Sol. A **150**, 763 (1995)
18. S. M. Kostritskii, D. Kip, E. Krätzig, Appl. Phys. B **65**, 517 (1997)
19. P. G. Suchoski, T. K. Findakly, F. G. Leonberger, Opt. Lett. **13**,1050 (1988)
20. A. Boudrioua, P. Moretti, J. C. Loulergue, J. Non-Cryst. Solids **187**, 443 (1995)
21. D. Kip, B. Kemper, I. Nee, R. Pankrath, P. Moretti, Appl. Phys. B **65**, 511 (1997)
22. A. Dazzi, P. Mathey, P. Lomprä, P. Jullien, P. Moretti, D. Rytz, J. Opt. Soc. Am. B **16**, 1915 (1999)
23. S. M. Kostritskii, P. Moretti, Appl. Phys. B **68**, 767 (1999)
24. T. Tamir: *Guided-wave Optoelectrics* (Springer, Berlin, Heidelberg 1988)
25. P. D. Townsend, Nucl. B **46**, 18 (1990)
26. P. D. Townsend, P. J. Chandler, L. Zhang: *Optical Effects of Ion Implantation* (Cambridge Univ. Press, Cambridge 1994)
27. A. Boudrioua, Ch. Bakouya, J. C. Loulergue, P. Moretti, K. Polgar, J. App. Phys. **89** (2001)
28. G. L. Destefanis, J. P. Gaillard, E. Ligeon, S. Valette, B. W. Farmery, P. D. Townsend, A. Perez, J. Appl. Phys. **50** 7898 (1979)
29. P. Moretti, P. Thevenard, K. Wirl, P. Hertel, Mater. Res. Soc. Symp. Proc. **244**, 323 (1992)
30. T. Pliska, D. Fluck, P. Günter, L. Beckers, C. Buchal, J. Opt. Soc. Am. B **15**, 628 (1998)
31. J. F. Ziegler, J. P. Biersack, U. Littmark: *The Stopping and Ranges of Ions in Solids* (Pergamon, New York 1988)
32. U. B. Ramabadran, H. E. Jackson, Appl. Phys. Lett. **58**, 672 (1991)
33. S. M. Kostritskii, P. Moretti, J. Mugnier: Advances in photorefractive materials, effects and devices, Trends Opt. Photon. Ser. (TOPS) **27** 361–367 (Opt. Soc. Am. 1999)
34. A. Alcazar de V., J. Rams, J. M. Cabrera, F. Agullo-Lopez, Appl. Phys. B. **68**, 989 (1999)

35. U. van Stevendaal, K. Buse, H. Malz, E. Krätzig, J. Opt. Soc. Am. B **15**, 2868 (1998)
36. K. Buse, Appl. Phys. B **64**, 361 (1997)
37. E. Chuang, D. Psaltis, Appl. Opt. **36**, 8445 (1997)

Two-Photon Optical Storage in Photorefractive Polymers in the Near-Infrared Spectral Range

Daniel Day[1], Min Gu[1], and Andrew Smallridge[2]

[1] Centre for Micro-Photonics, School of Biophysical Sciences and Electrical Engineering, Swinburne University of Technology
PO Box 218, Hawthorn, Victoria 3122, Australia
[2] School of Life Sciences and Technology, Victoria University of Technology
PO Box 14428 MCMC, Melbourne, Victoria 8001, Australia
{dday,mgu}@groupwise.swin.edu.au

Abstract. We report the use of a polymer-based photorefractive material for three-dimensional bit optical data storage using near-infrared illumination. The research was conducted using photorefractive materials that were fabricated in two polymer matrices: poly(N-vinylcarbazole) (PVK) and poly(Methyl Methacrylate) (PMMA). The recording samples also consisted of the following compounds in various proportions: 2,5-dimethyl-4-(p-nitrophenylazo)anisole (DMNPAA), 2,4,7-trinitro-9-fluorenone (TNF) and N-ethylcarbazole (ECZ). Two-photon excitation was used as the recording mechanism to achieve rewritable bit data storage in a photorefractive polymer. As a result of two-photon excitation, the quadratic dependence of the excitation on the incident intensity produces an excitation volume that is confined to the focal region in both the transverse and axial directions. The use of ultrashort pulsed lasers, while effective, is not a practical solution for an optical data storage system. This research demonstrates the ability to produce three-dimensional rewritable bit data storage using continuous-wave illumination. Using this technology it has been possible to achieve a density of 88 Gbits/cm^3, which in the future could be increased to 3.5 Tbits/cm^3.

1 Three-Dimensional Bit Optical Data Storage in a Photorefractive Polymer

Multi-layered (or three-dimensional) optical memories have increasingly become a field of interest in the development of high-density optical data storage devices [1]. Systems that utilize multiple-layer recording can achieve a recording density from 100 to 10000 times higher than that in conventional optical data storage devices. Three-dimensional bit optical data storage has the ability to reach a density of Tbits/cm^3 [2].

Multi-layered bit storage is possible with the use of two-photon excitation as summarized in Chap. 1 of this book. The probability of two-photon excitation is proportional to the square of the incident intensity; this results in excitation within a small region of the focus spot of the recording objective.

P. Boffi, P. Piccinin, M. C. Ubaldi (Eds.): Infrared Holography for Optical Communications, Topics Appl. Phys. **86**, 75–90 (2003)

The focus spot can therefore be positioned at any depth within the volume recording medium, which leads to effective recording of layers of information. As a result of the quadratic relationship between the incident intensity and the material excitation, information layers can be recorded at small layer separations with little cross talk between the layers. Another advantage of two-photon excitation is the use of infrared illumination, which results in the reduction of scattering and enables the recording of layers at a deep depth in a thick material.

While pulsed beam illumination is the most efficient method of recording under two-photon excitation, it will be demonstrated that rewritable bit storage in the photorefractive polymer can be achieved using continuous wave illumination.

The aim of this chapter is to demonstrate the feasibility of the photorefractive polymer for three-dimensional bit optical data storage. The recording and reading methods will be discussed as well as the characterization of the recorded bit sizes.

2 Experimental Recording System

A schematic diagram of the experimental recording system used to record information within three dimensions of a volume recording material is illustrated in Fig. 1.

To control the logic state of the recorded information a mechanical shutter was used to block the beam. In these experiments an NM Laser (NM Laser Product Inc., Sunnyvale, CA) shutter with a minimum rise and fall time of 2 ms was used. This limited the speed of the recording to 500 bits per second (bps). Typically the shutter was operated with a rise and fall time of 15 ms, to avoid any inaccuracy in the exposure time. After the shutter, the beam was spatially filtered and collimated. Using a 10× objective (O_1) with numerical aperture (NA) of 0.25, the laser beam was focused through a pinhole (P_1), typically $10 - 30\,\mu m$ in diameter. The light collected from the pinhole was then collimated with lens L_1. The focal length of L_1 and therefore the collimated beam diameter were dictated by the size of the back aperture of the recording objective (O_2). Following L_1 was a variable aperture (A) and a 50/50 beam splitter (BS_1). The initial beam-steering mirrors (M_1, M_2) and the two beam splitters were coated for ultra-short pulses, to reduce the amount of pulse broadening throughout the system.

An increase in the pulse width decreases the peak power within the focus of the objective, which therefore decreases the efficiency of two-photon excitation. The relationship between the rate of two-photon excitation and the pulse width will be discussed in Sect. 6, which considers continuous-wave two-photon excitation in the recording process.

The recording sample was mounted in a computer-controlled translation stage. Two recording stages were used throughout this work. The first

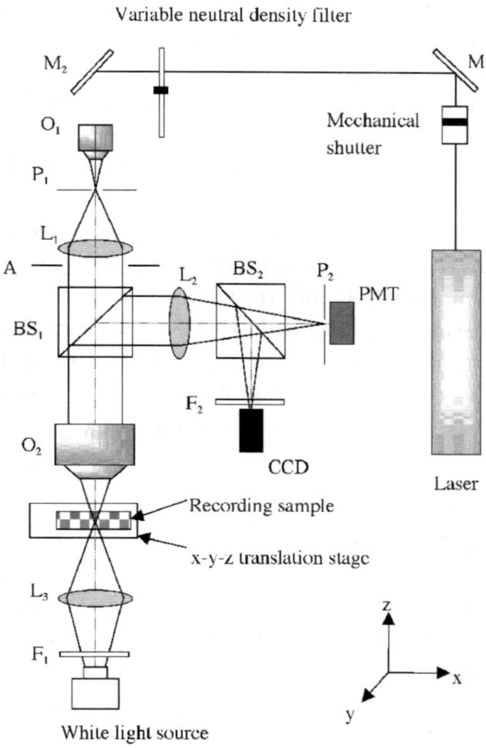

Fig. 1. Schematic diagram of the recording system

was a Melles Griot Nanoblock system, and the second was a Melles Griot NanoMover. The Nanoblock consisted of a 200 × 200 µm x–y piezoelectric stage and a 200 µm z piezoelectric stage. The displacement of the stage was controlled by an analogue voltage signal from a computer sent to the drive box. The resolution of each axis in the Nanoblock was 50 nm. The NanoMover translation stage has three individual axes, each with a travel of 25 mm. To achieve the longer travel, stepper actuators were used in the NanoMover, which results in a slightly reduced accuracy of 100 nm and a repeatability of 200 nm, compared to the Nanoblock. GPIB communication was used to control the NanoMover.

The recording program for both systems was written using the software LabVIEW. The program was designed to accept an input file made up of 1's (bit) and 0's (no bit), corresponding to where a bit was to be recorded.

In order to determine the position of the focus before recording the sample was scanned in the z (axial) direction. Scanning the sample in the z direction produced an axial response. The scanned signal from the surface of the sample was collected by the objective (O_2) and reflected from BS_1 through another lens (L_2) which focused the light into a pinhole (P_2) mounted in front of a detector, typically a photomultiplier tube (PMT). Once the surface of the sample was determined, the focus position could be accurately translated throughout the sample to record multiple layers.

To view the recording process in real time, a white light source was focused by lens L_3 onto the back of the recording sample. The light was then collected by the objective, O_2, and reflected by BS_1 towards the PMT. Between L_2 and the detection pinhole another beam splitter was placed to reflect the light onto a charge coupled device (CCD). The image from the CCD was collected by a frame grabber and displayed in real time on a computer monitor. A long-pass filter (F_1), with a cutoff wavelength of 600 nm, was placed after the white light source to prevent any unnecessary erasure of the recorded information, as the recording sample has strong absorption in the ultra-violet to visible wavelength region. A short-pass filter (F_2) with a cutoff wavelength of 750 nm was placed in front of the CCD, to reduce the chances of damaging the CCD array from the focused laser light.

3 Experimental Reading System

The change in refractive index created by the recording process in the photorefractive polymer is typically less than a 1% change. Such a small change can only be measured by an optical system that is sensitive to the phase change of a reading beam. The two methods that were used in this research were transmission and differential interference contrast (DIC) imaging.

3.1 Transmission Reading

The transmission system is the simplest method for detecting phase changes along the optical axis of a material. The use of a light source illuminating the sample and being collected by the CCD represents a transmission system. However, the image collected by the CCD camera does not have high enough resolution to read the recorded information for the purposes of this work.

For this project an Olympus FluoView scanning microscope was used to read the recorded information. The FluoView microscope is designed for single-photon fluorescence, and as such has a krypton:argon laser coupled to it for ultraviolet (UV)–blue excitation at wavelengths of 488 and 564 nm.

As these wavelengths are within the absorption band of the photorefractive polymer, they cannot be used as they will erase the recorded information. Instead a helium–neon laser (632.8 nm) is coupled to the microscope. To improve the performance, the dichroic beamsplitter designed to reflect the laser and transmit the fluorescence (for normal operation) is replaced with a standard 50/50 beamsplitter.

Although able to detect the changes in refractive index of a recorded bit, the transmission system has poor axial resolution, and therefore limits the minimum distance between recorded layers.

3.2 Differential Interference Contrast Reading

Nomarski differential interference contrast (DIC) imaging [3] has become a widely used tool in microscopy for retrieving phase information from a sample. A couple of key features of DIC microscopy result in its popularity compared with the older methods of phase imaging, based on Zernike phase contrast systems. The first feature is the ability to obtain both phase and intensity and display them in the form of a shadowed boundary at the point of increasing or decreasing phase. The amount of highlighting produced at the boundary can be varied by changing the phase delay introduced into the system. The second feature is the apparent optical-sectioning property compared with other conventional microscope imaging techniques [4]. However, it should be noted that this only applies to the higher spatial frequency components of the object and cannot be compared with the optical-sectioning property of confocal microscopy.

The image from DIC microscopy is produced by the difference in the amplitudes of two images, which are displaced laterally and have been phase shifted with respect to each other [5]. A plane polarized laser beam is split into two components by a Wollaston prism. The two waves, e- and o- waves, become spatially separated and are focused into the sample. While passing through the sample, the two beams will experience different amounts of phase shift if they pass through different regions of refractive index. The beams are then collected and recombined with another Wollaston prism before passing through an analyser and being detected. An image is formed by scanning the sample through the focus position of both beams.

To optimize the sensitivity of the system and the contrast of the resulting image, control over the amount of lateral shear and phase shift of the two components is required. The lateral shear is a function that is designed into the Wollaston prism, while the phase shift is adjusted by lateral displacement of the second Wollaston prism analyser. It is also possible to alter the relative strength of the two beams by rotating the two polarizers.

As will be shown in Sects. 4 and 6 using DIC microscopy to image the small changes in refractive index associated with a recorded bit is an effective method, though there are a number of disadvantages. The poor resolution in the axial direction is a limiting factor in the density of recorded information, and the optical system required is not easily incorporated into a compact device.

4 Pulsed Beam Illumination

To achieve high-efficiency two-photon excitation, an ultrashort pulsed laser is needed. In these experiments the laser used was a Spectra Physics Tsunami, pumped by a 10 W Millennia. The Tsunami is an ultrashort pulsed Ti:sapphire laser, which has a wavelength range of 700–1000 nm. Within the wavelength range it produces ultrashort pulses of ≈ 70 fs at a repetition rate of 82 MHz. The laser can also be operated in a continuous wave (CW) mode.

4.1 Multi-layered Data Storage

With two-photon excitation for recording, the first demonstration of three-dimensional bit optical data storage in photorefractive polymers was achieved [6]. The photorefractive material used consisted of nonlinear chromophore 2,5-dimethyl-4-(p-nitrophenylazo)anisole (DMNPAA), the photosensitive compound 2,4,7-trinitro-9-fluorenone (TNF) and the plasiticizer N-ethylcarbazole (ECZ), all doped into the polymer host poly(N-vinylcarbazole) (PVK). The concentration of the materials DMNPAA:PVK:ECZ:TNF was 10:73:16:1 as a percentage of the total weight. Owing to difficulties in manufacturing a homogenous sample with high overall concentration of dopants, the samples were manufactured with a thickness ranging from 100 to 200 μm.

Figure 2 demonstrates the ability to record multiple layers of information within the volume of a photorefractive polymer under pulsed two-photon excitation. The wavelength used for excitation was 800 nm, with an average power of 7.5 mW in the focus of a Zeiss Fluar objective with a numerical aperture and a magnification factor of 0.75 and 20, respectively. The objective operated at an infinite tube length and was corrected for a 170 μm thick cover slip. The exposure time for each bit was 20 ms. The spacing between bits in each plane was 3.2 mm, while the spacing between layers was 20 mm. With this bit and layer spacing the density of the recorded information was 5 Gbits/ cm^3. In this case the layer spacing was large enough to prevent cross-talk from the neighbouring layers.

For reading the change in refractive index caused by the two-photon photorefractive process we employed an Olympus FluoView scanning microscope and used it in a DIC mode. A He–Ne laser of wavelength 632.8 nm was coupled in the FluoView microscope for reading the recorded information, as the wavelength of 632.8 nm is on the edge of the absorption band and causes minimal erasure to the recorded information [6]. The power within the focal region of an Olympus UPlanApo objective was less than 2 mW. The numerical aperture and magnification factor of the objective were 0.7 and 20, respectively. The objective operated at an infinite tube length and was corrected for a 170 μm thick cover slip.

The values for recording power and exposure time were chosen so as to create a large-enough change in refractive index to provide good contrast in the DIC image and to maintain the ability to erase, while a high density of information was maintained. A further discussion on the relationship between recording power, exposure time and numerical aperture is given in Sect. 5.

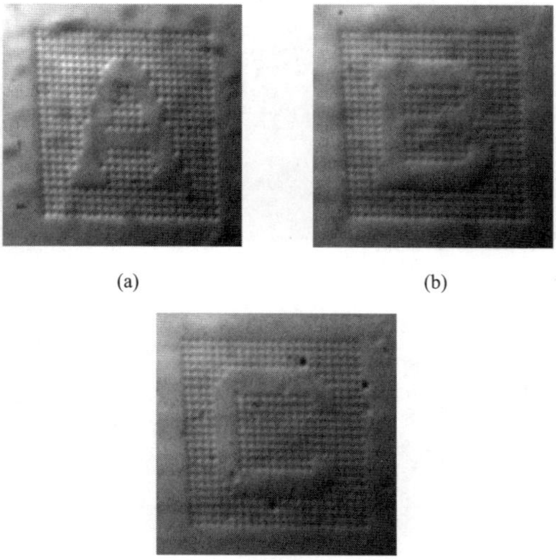

(a) (b) (c)

Fig. 2. Recorded 24×24 bit patterns at different depths in the photorefractive polymer under two-photon excitation. The spacing between adjacent layers is 20 mm, and the bit separation is 3.2 mm. (**a**) the first layer including the letter A, (**b**) the second layer including the letter B, and (**c**) the third layer including the letter C

4.2 Rewritable Data Storage

The principle of the photorefractive effect is that a nonuniform illumination will lead to a position-dependent space-charge field, which in turn modulates the refractive index of an electro-optic material. If the charges are redistributed uniformly, then the information based on the change in refractive index is erased. Figure 3 demonstrates that due to the photorefractive effect, an array of bits can be written, erased and rewritten under two-photon excitation for recording and ultraviolet illumination for erasing.

The parameters for the recording process are the same as used for Fig. 2. The background in Fig. 3 has increased shadowing as a result of an increase in the lateral shear. To erase the information the sample was illuminated with the ultraviolet line of the mercury lamp in the FluoView microscope. A typical time for complete erasure of the information was 1–2 s. The rewritten pattern (see Fig. 3c) was obtained using the same recording parameters as mentioned above.

Using erasable materials may lead to an issue regarding the stability of the recorded information. Figure 4 shows the deterioration of the recorded information after being read 1000 times. The contrast of the recorded bits in Fig. 4b are 50% of that in Fig. 1a. This result illustrates that there is weak erasure of the information due to the slight absorption of the light beam used for reading.

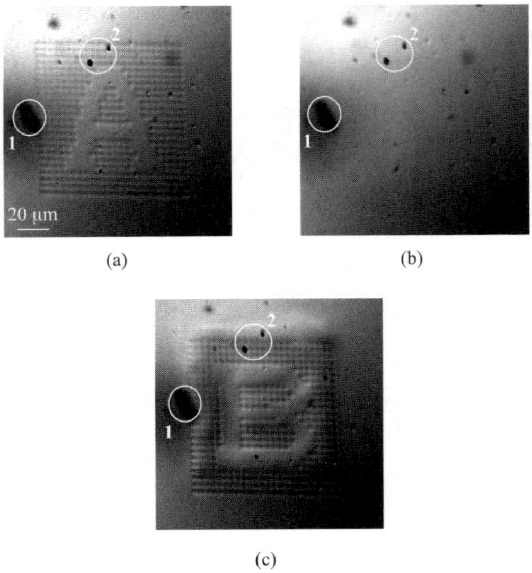

Fig. 3. Demonstration of writing, erasing and rewriting in the same area. (**a**) letter A is recorded, (**b**) letter A is erased after being exposed to UV illumination for 1-2 s, and (**c**) letter B is recorded in the same area. The marked artifacts 1 and 2 indicate that the images are in the same area

There is also strong absorption of sunlight and fluorescent lights, which result in erasing the information. If kept in the dark, the information will maintain its contrast for several hours, at which point thermal relaxation destroys the quality of the information. Thermal relaxation in this particular material is strong due to the reduction in the glass transition T_g temperature by the introduction of the plastisizer. A better stability could be achieved by reducing the absorption band and increasing the glass transition temperature of the sample.

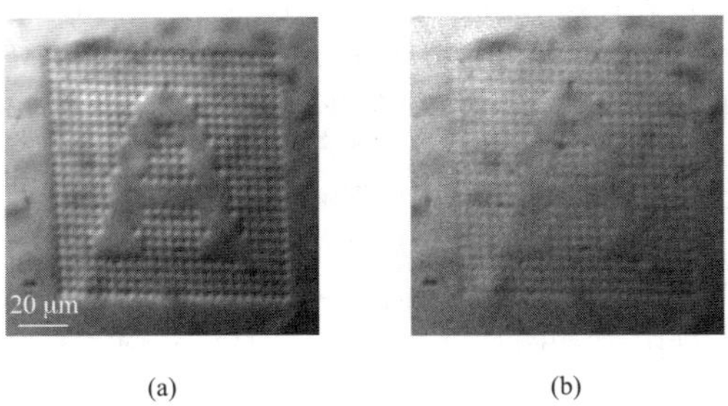

Fig. 4. Images of 24×24 bit patterns recorded by two-photon excitation in a photorefractive polymer. (**a**) letter A after first reading, and (**b**) letter A after reading 1000 times

5 Bit Characterisation

In order to understand the relationship between the recording parameters and the resulting change in refractive index, a characterization of the recorded bits was completed. From a practical point of view a low numerical aperture objective was used in the recording process, as there is less spherical aberration involved with the smaller focusing angles. This was then compared with the performance of an oil immersion objective with high numerical aperture.

Figure 5 shows the relationship of the bit size to the recording power, exposure time and recording depth. In recording, an Olympus ultralong working distance (ULWD) objective with a numerical aperture and a magnification factor of 0.8 and 20, respectively, was used. For these experiments the recording laser was operated in a pulsed mode at a wavelength of 800 nm.

The bits were read using the Olympus FluoView microscope operated in transmission, with an Olympus PlanApo oil immersion objective that had a numerical aperture and a magnification factor of 1.4 and 604, respectively. The bit spacing was increased to 7.8 μm to prevent any interference on the bit size from neighbouring bits. The exposure time was maintained at 25 ms while the recording power was varied, or the power was kept at 25 mW while the exposure time was altered. For recording at different depths the power and exposure time were 25 mW and 25 ms, respectively.

From Fig. 5, it can be seen that an increase in power, exposure time and depth results in an increase in the bit size. In Figs. 5a and b there is a region at the highest power and longest exposure time where the bit size decreases. It is at this point where a micro-cavity is formed. However, the change in bit size as a function of depth is a result of spherical aberration introduced from the mismatch in refractive indices [7] between the immersion (air $n = 1.0$) and recording (polymer $n = 1.49$) media.

Figure 6 shows the relationship of the spot size to the recording power, exposure time and recording depth for an oil immersion high numerical aperture objective. The objective used was a Zeiss Fluar objective with numerical aperture and a magnification factor of 1.3 and 40, respectively. To demonstrate the relationship of the bit size to power the exposure time was maintained at 25 ms. To illustrate the relationship between bit size and exposure time the power was kept at 14 mW. The power and exposure time were kept at 14 mW and 25 ms, respectively, when the bit size versus depth relation was measured. The reading conditions are the same as described previously in this section.

The same characteristics, as those shown in Fig. 5, can be seen in Fig. 6 using the high numerical aperture objective. However, a comparison of the corresponding graphs shows that using the higher numerical aperture objective results in a smaller bit size under almost all recording conditions except for those in Fig. 5c and Fig. 6c. This feature results from the smaller focal region produced by the higher numerical aperture objective.

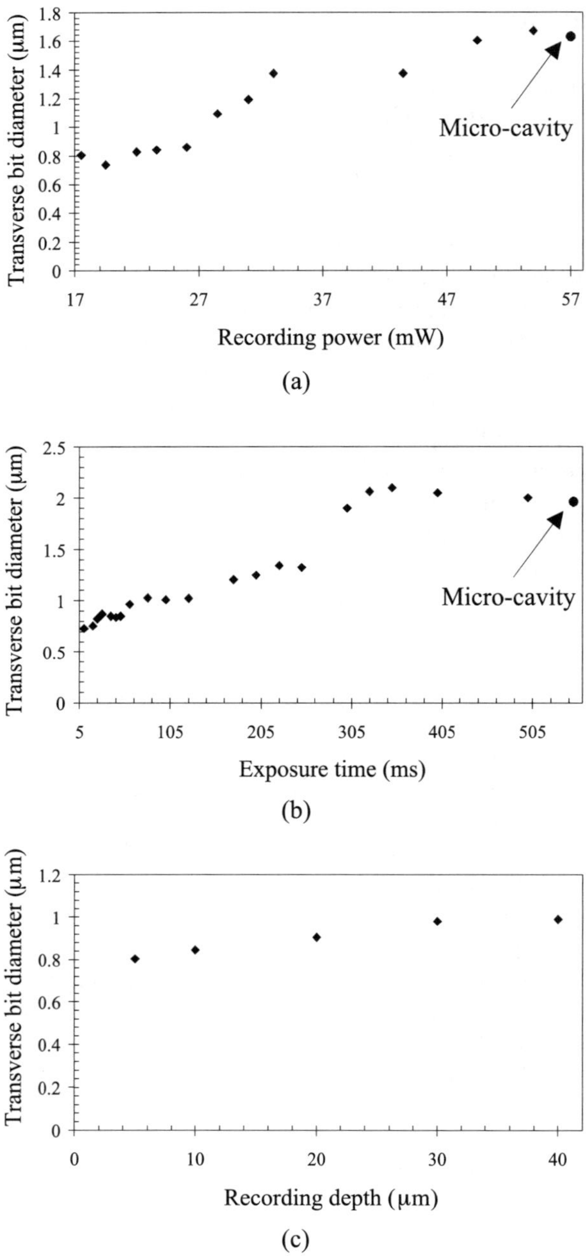

Fig. 5. Relationship between bit size and (**a**) power, (**b**) exposure time and (**c**) recording depth, for a recording objective of numerical aperture 0.8. The points marked by a diamond (♦) indicate erasable data storage, and the points marked by a *circle* (•) are conditions under which micro-cavities are formed

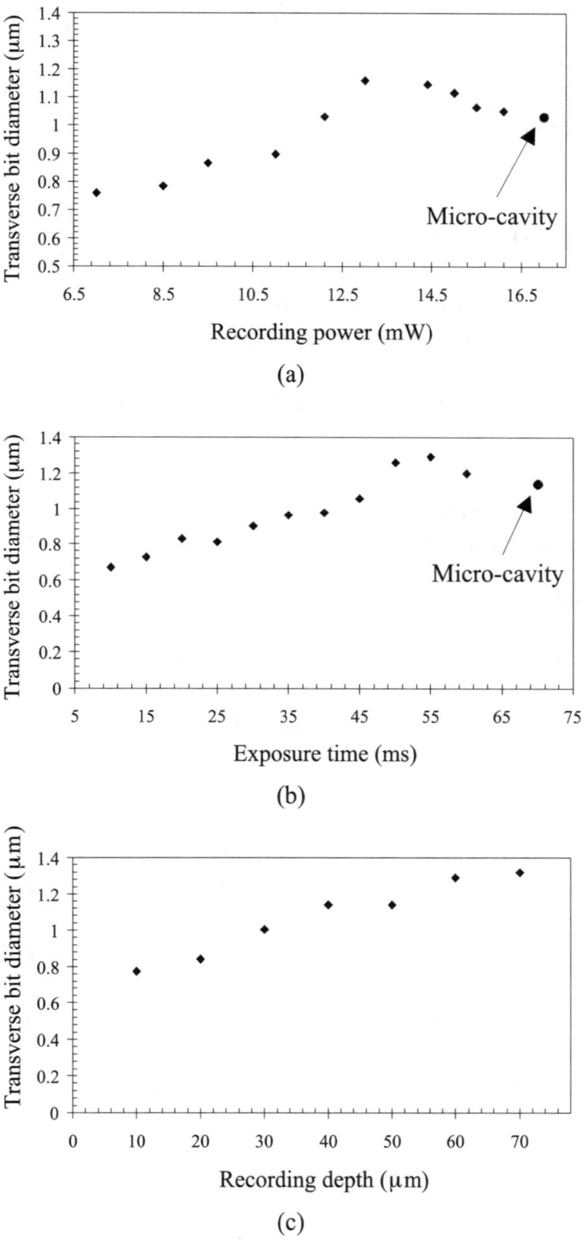

Fig. 6. Relationship between bit size and (**a**) power, (**b**) exposure time and (**c**) recording depth, for a recording objective of numerical aperture 1.3. The points marked by a *diamond* (♦) indicate erasable data storage, and the points marked by a *circle* (•) are conditions under which micro-cavities are formed

Figure 6c further illustrates the effect of spherical aberration as a reslt of the mismatched refractive indices. Although the immersion oil has a refractive index of 1.513, which is close to that of the polymer (1.49), there is still a pronounced effect of spherical aberration due to the use of a high numerical aperture objective.

6 Continuous-Wave Illumination

Two-photon excitation is the important recording method for three-dimensional optical data storage. However, it requires the use of an ultrashort pulsed laser for efficient excitation as described in Chap. 1, Sect. 2.2, which is an impractical method. A better alternative would be to use an inexpensive laser diode.

6.1 Requirements for Two-Photon Excitation with Continuous-Wave Illumination

An early report on continuous-wave illumination for two-photon excitation by *Hänninen* et al. [8] demonstrated the ability to produce fluorescence from biological samples. This report also demonstrated a relationship between fluorescence intensity and pulse width. Although they considered a square pulse function, they found considerable agreement between their expression and the experimental results. It was soon discovered [9] that the high average powers required to achieve an efficient two-photon excitation rate was quite likely damaging the biological tissue.

According to *Denk* et al. [10] the following is a more accurate relationship between the illumination parameters and the two-photon absorption rate, ϕ_{fl}:

$$\phi_{\mathrm{fl}} = \frac{0.675}{\tau f} \frac{\sigma_2 P_{\mathrm{AVE}}^2}{(h\omega A)^2} , \tag{1}$$

where the factor 0.675 holds for a secant-squared pulse shape with a pulse width and a repetition rate τ and f, respectively; σ_2 is the two-photon absorption cross-section, $h\omega$ is the photon energy and A is the focal area. The relationship in (1) leads to

$$P_{\mathrm{CW}} = P_{\mathrm{AVE}} \sqrt{\frac{0.675}{\tau f}} \tag{2}$$

for achieving the same two-photon absorption rate under pulsed and continuous-wave illumination. For a repetition rate of 82 MHz and a pulse width of 200 fs in the focus the average power for recording needs to be increased approximately by a factor of 200.

6.2 Continuous-Wave Multi-Layered Data Storage

Continuous wave two-photon excitation has been previously demonstrated for bit optical data storage in a photobleaching polymer [11] as well as holographic data storage in a photorefractive crystal [12].

The material used in Sect. 4.1 for multi-layered data storage was based on the polymer host PVK. However, it was found that under the high average powers required for continuous-wave two-photon excitation PVK can be melted. The polymer poly(methyl methacrylate) (PMMA) was chosen due to its higher melting point [13]. The recording sample consisted of DMN-PAA:PMMA:TNF:ECZ whose percentages of total weight 30:53:16:1. Due to the high concentration of dopants the sample can only be fabricated to 100–200 μm in thickness without reducing significantly the optical quality of the sample.

Figure 7 demonstrates the ability to record three-dimensional optical data storage in the photorefractive polymer under continuous-wave two-photon excitation. For these experiments a Zeiss NeoFluar oil immersion objective with a numerical aperture of 1.3 and a magnification factor of 40 was used. The higher numerical aperture results in an increased power density of 34 times that of the lower numerical aperture objective used under pulsed excitation.

The wavelength used for recording was 800 nm with an average power of 77 mW in the focus and an exposure time of 10 ms. A bit spacing of 6.75 mm and a layer spacing of 20 μm produced a density of 1.1 Gbits/ cm^3. Owing to the difficulty in manufacturing thick recording samples and the large layer spacing only a few layers could be recorded.

The bits in Figs. 7 and 8 appear to be bumps rather than pits as shown in Figs. 2 and 3, the reason for the difference is due to the position of the focus above or below the bits and the amount of shearing introduced by DIC microscopy.

The images in Fig. 7 were produced by scanning the sample in the Olympus FluoView microscope in a DIC mode. The laser source used was a He–Ne laser of wavelength 632.8 nm and was focused through an Olympus UPlanApo objective with a numerical aperture of 0.7 and a magnification factor of 20.

6.3 Continuous Wave Rewritable Data Storage

To confirm that using the high average power for recording under continuous-wave two-photon excitation creates a change in refractive index via the photorefractive effect, the rewritable feature was demonstrated.

Figure 8 shows the ability to record, erase and record information within the same region of the material under continuous-wave two-photon excitation for recording and ultraviolet illumination for erasing. The wavelength for recording was 800 nm, with a recording average power of 300 mW in the focal region of an ULWD objective with a numerical aperture of 0.75 and

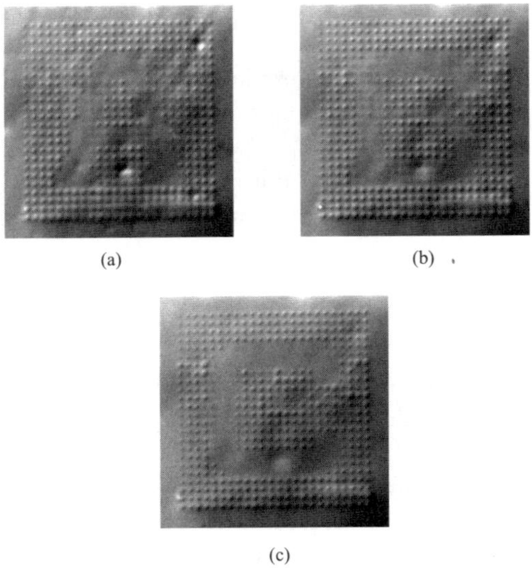

Fig. 7. Three-dimensional bit optical data storage in a photorefractive polymer under continuous-wave two-photon excitation. (**a**) the first layer including the letter A, (**b**) the second layer including the letter B, and (**c**) the third layer including the letter C

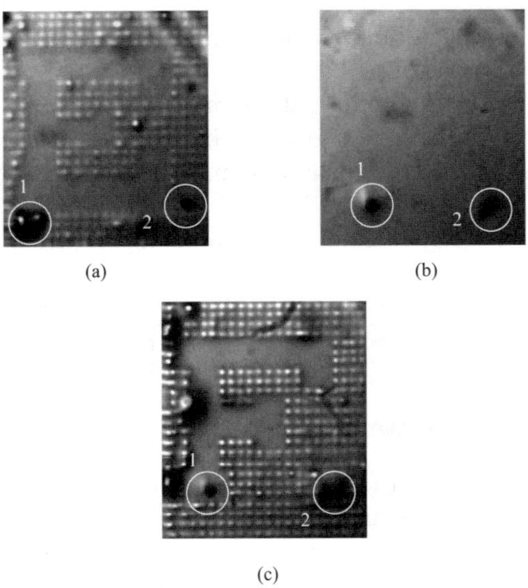

Fig. 8. Rewritable bit optical data storage in a photorefractive polymer under continuous-wave two-photon excitation. (**a**) the letter E is recorded. (**b**) the letter E is erased after illuminating the same region with UV light. (**c**) the letter F is recorded into the same region as indicated by the artifacts 1 and 2

a magnification factor of 40. An exposure time of 2 ms was used. The bit spacing was 5.6 mm.

The images were read using the Olympus FluoView microscope in a DIC mode. The reading wavelength was 632.8 nm and was focused by an Olympus UPlanApo objective with a numerical aperture and a magnification factor of 0.7 and 20, respectively.

The recorded bits in Fig. 8 are significantly stronger than that observed in Fig. 7 for the multi-layered recording. This is due to the proximity of the surface. It was discovered that it was easy to damage the material when recording within 10 μm of the surface with high power. When a laser beam is focused near the surface under high recording powers, there is sufficient heating to ablate the recording medium.

The stability of the recorded information under continuous-wave two-photon excitation is approximately the same as that under pulsed illumination, as they both use the photorefractive effect to create the recorded bit.

7 Conclusions

The work covered in this chapter represents a detailed investigation into three-dimensional bit optical data storage in a photorefractive polymer. Key areas of research include the fabrication of a photorefractive polymer and a multi-layered erasable/rewritable optical memory under two-photon excitation with nearinfrared pulsed and continuous-wave illumination.

A photorefractive polymer based on the compounds 2,5-dimethyl-4-(p-nirtophenylazo)anisole (DMNPAA), 2,4,7-trinitro-9-fluorenone (TNF) and N-ethylcarbazole (ECZ) in either poly(N-vinylcarbazole) (PVK) or poly(methyl methacrylate) (PMMA) as the host matrix was fabricated and used as the recording medium for this work.

The creation of a highly localized modulation of the refractive index under two-photon excitation in the photorefractive polymer provided the ability to record multi-layered information. As the photorefractive effect is a reversible process, rewritable bit optical data storage was also demonstrated. While two-photon excitation provided the means to record multi-layered information, it requires a high photon density within the focal spot for efficient excitation. Typically an ultrashort pulsed laser is used for this process; however, such a laser is impractical. Continuous wave two-photon excitation was applied to record rewritable multi-layered information in a photorefractive polymer.

In conclusion, this research shows the ability to record multi-layered rewritable information in a photorefractive polymer under near-infrared two-photon pulsed and continuous wave illumination. It has also been demonstrated that with the performance of the current recording and reading systems it is possible to achieve a density of greater than 88 Gbits/ cm^3. If the effect of spherical aberration caused by the mismatch in refractive indices is compensated [7,14], a data density of Tbits/ cm^3 could be possible in the near future.

References

1. S. Kawata: Three-dimensional optical data photorefractive bit-oriented digital memory, in F. T. S. Yu, S. Yu (Eds.): *Photorefractive Optics* (Academic, San Diego 2000), pp. 277–306
2. J. H. Strickler, W. W. Webb: Three-dimensional optical data storage in refractive media by two-photon point excitation, Opt. Lett. **16**, 1780–1782 (1991)
3. G. Normaski: Microinterférométrie differential et ondes polarisés, J. Phys. Radium **16**, 9–135 (1955)
4. C. Preza, D. Snyder, J. A. Conchello: Theoretical development and experimental evaluation of imaging models for differential-interference-contrast microscopy, J. Opt. Soc. Am. A **16**, 2185–2199 (1999)
5. C. J. Cogswell, C. J. R. Sheppard: Confocal differential interference contrast (DIC) microscopy: including a theoretical analysis of conventional and confocal DIC imaging, J. Micros. **165**, 81–101 (1992)
6. D. Day, M. Gu, A. Smallridge: Use of two-photon excitation for erasable-rewritable three-dimensional bit optical data storage in a photorefractive polymer, Opt. Lett. **24**, 948–950 (1999)
7. D. Day, M. Gu: Effect of refractive-index mismatch on three-dimensional optical data storage density in a two-photon bleaching polymer, Appl. Opt. **37**, 6299–6304 (1998)
8. P. E. Hänninen, E. Soini, S. W. Hell: Continuous wave excitation two-photon fluorescence microscopy, J. Micros. **176**, 222–225 (1994)
9. K. König, H. Liang, M. W. Berns, B. J. Tromberg: Cell damage by near-IR microbeams, Nature **377**, 20–21 (1995)
10. W. Denk, J. H. Strickler, W. W. Webb: Two-photon laser scanning fluorescence microscopy, Science **248**, 129–131 (1990)
11. M. Gu, D. Day: Use of continuous wave illumination for two-photon three-dimensional optical data storage in a photobleaching polymer, Opt. Lett. **24**, 288–290 (1999)
12. J. Imbrock, D. Kip, E. Krätzig: Nonvolatile holographic storage in iron-doped lithium tantalate with continuous-wave laser light , Opt. Lett. **24**, 1302–1304 (1999)
13. D. Day, M. Gu, A. Smallridge: Rewritable 3D bit optical data storage in a PMMA-based photorefractive polymer, Adv. Mater. **13**, 1005–1007 (2001)
14. M. Gu, D. Day, O. Nakamura, S. Kawata: Three-dimensional coherent transfer function for reflection confocal microscopy in the presence of refractive index mismatch, J. Opt. Soc. Am. A **18**, 2002–2008 (2001)

Long-Lifetime Photorefractive Holographic Devices via Thermal Fixing Methods

Mercedes Carrascosa, José M. Cabrera, and Fernando Agulló-López

Departamento Fisica de Materiales C-IV, Universidad Autónoma de Madrid
Canto Blanco, Madrid 28049, Spain
m.carrasocosa@uam.es

Abstract. This contribution reviews the present knowledge on photorefractive thermal fixing, considering phenomena and their applications to long-lifetime holographic devices. Most attention is devoted to $LiNbO_3$, where a coherent and clear picture has recently emerged and most applications are implemented. Discussions an the physical basis for thermal fixing as well as an the standard mathematical model are presented. Relevant experimental results an thermal fixing are described and the capability of the model to reliably interpret them is emphasized. In addition, either bulk or waveguide devices of potential interest in telecommunications are revised and discussed.

1 Introduction

Since its discovery in 1966 [1], the photorefractive effect has given rise to a large variety of experiments and applications [2,3,4,5]. Good examples are information storage, real-time holographic interferometry, beam modulators and deflectors, extremely narrow-band interference filters, optical (image and signal) amplifiers, optical phase conjugators, optical information processing and computing, material characterization, etc. Many of these applications, including telecom applications, require a long lifetime of the patterns recorded by the photorefractive effect, so a considerable effort has been devoted to the subject. The need for this effort is evident as, apart from other reasons, photorefraction is a reversible effect in which the reading light beam continuously erases the recorded pattern.

Among the various techniques used for increasing the duration of a photorefractive pattern, the method first reported by *Amodei* and *Staebler* in 1972 [6,7] and known as *thermal fixing* continues to be the most popular one. In fact, thermal fixing has produced the first marketed photorefractive device consisting of a very narrow bandwidth interference filter [8,9]. The present chapter focuses on the physical basis for thermal fixing (including important experiments and theoretical developments), major recent advances to increase the storage lifetime, and relevant applications.

P. Boffi, P. Piccinin, M. C. Ubaldi (Eds.): Infrared Holography for Optical Communications,
Topics Appl. Phys. **86**, 91–110 (2003)

2 The Photorefractive Effect

Photorefractive phenomena were first discovered in $LiNbO_3$ and then in a wide variety of electro-optic/photoconductor materials. Detailed descriptions of photorefractive phenomena, materials and applications are given in a number of books [2,3,4,5]. Photorefractive recording of a certain light pattern consists of three basic steps (Fig. 1). (a) photoexcitation of trapped electrons (from some electron traps such as Fe^{2+}); (b) migration of free carriers from bright regions into dark regions driven by some transport mechanism (i. e. externally applied electric field, bulk photovoltaic effect, diffusion in a carrier concentration gradient); (c) electron retrapping at some electron traps (such as Fe^{3+}) in dark regions. (d) At the end of these steps, the electric field associated with the trapped charge distribution gives rise to a refractive index pattern (Fig. 1e) via the linear electro-optic (Pockels) effect that matches the original light pattern.

The physical model just described predicts that any photoconductor/electro-optic material, with appropriate electron traps, is in principle a candidate material to show photorefractive phenomena. As a matter of fact, photorefractive phenomena have also been observed in quadratic electro-optic (Kerr) materials, although in general the efficiency is substantially lower. Among the various photorefractive active impurities, the Fe^{2+}–Fe^{3+} system is the most efficient in $LiNbO_3$, as well as in a number of other crystals ($LiTaO_3$, $BaTiO_3$, $KNbO_3$, ...). In $LiNbO_3$, the photorefractive efficiency can be substantially reduced by hydrogen doping as well as by some types of divalent cation doping, among which Mg^{2+} [10,11] is the most popular one. Most of the photorefractive applications [2,12,13,14] are based on the

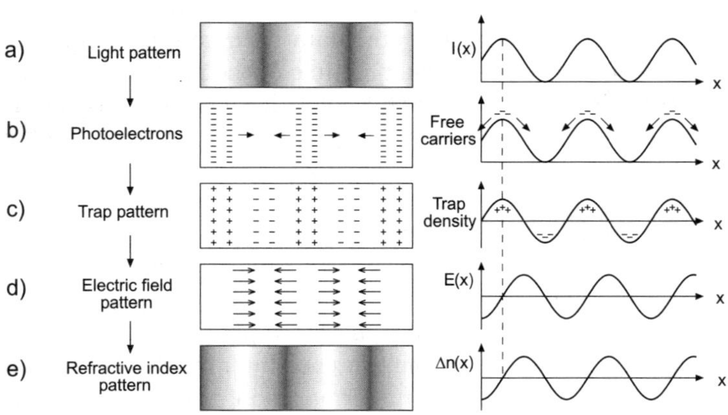

Fig. 1. Sketch of the basic steps of photorefractive recording. (**a**) The sample is illuminating with a sinusoidal light pattern. (**b**) Generated photoelectron pattern from filled traps. (**c**) Transport and retrapping give rise to an ionized trap pattern. (**d**) The associated space electric field pattern ($\pi/2$ phase shifted). (**e**) Refractive index pattern induced by the linear electrooptic effect

production of phase volume holograms (for information storage and optical interconnections) and on nonlinear wave mixing (for image processing and optical phase conjugation).

3 Photorefractive Fixing Techniques

A pattern recorded in a photorefractive material, such as the one shown in Fig. 1d, is partially erased on every readout. The main reason is the redistribution of trapped charges induced by the homogeneous photoionization of the reading beam. This is a major drawback in applications in which a permanent pattern is needed, so several methods have been developed to partially overcome the problem. Some methods are based on using a reading wavelength with very low photorefractive efficiency, so optical erasing is negligible (nondestructive readout). This can be achieved by reading at a different wavelength (for example infrared) from the recording one (for example green) and is usually called the *two-wavelength technique*. An alternative procedure is so-called *two-photon recording*. This method consists of illuminating the crystal with an additional light source which is not present during reading. The additional light acts as an activator during recording and its absence during reading prevents pattern degradation. Both subjects are considered in separate chapters in this book.

Fixing methods are based on producing a light-insensitive replica of the electronic pattern via some structural property of the crystal, such as the spontaneous polarization in ferroelectric crystals (*electrical fixing*), or some mobile ionic charges, such as protons in $LiNbO_3$ and other similar oxides (*thermal fixing*). The following sections are devoted to thermal fixing, which is the best known and extended method, although electrical fixing has even been applied to channel waveguides in strontium-barium niobate crystals [15].

4 Physical Model for Thermal Fixing

As $LiNbO_3$ remains the most promising crystal for storage applications, particularly via thermal fixing, and the majority of work has been done on this material, we will mainly refer the discussion to $LiNbO_3$.

4.1 Standard Model for $LiNbO_3$

Amodei and *Staebler* [6,7] first reported the thermal fixing process as the generation of a replica of a light-erasable trapped charge pattern, such as that sketched in Fig. 1c, into a light-insensitive (fixed) matching ionic pattern. It is now generally accepted that in $LiNbO_3$ [16,17,18,19] thermal fixing is produced by the migration of protons which are present in common as-grown crystals. In $LiNbO_3$ protons are mobile at temperatures above $\approx 70\,°C$, whilst trapped electrons are thermally excited at a much lower rate between this temperature and ≈ 180 °C in non-reduced $LiNbO_3$. (In reduced samples, shallow electron traps appear whose detrapping temperature is much lower [20].) Between these two temperatures, typically around 150 °C, protons can migrate driven by the space-charge field created by the trapped electrons, the process being much faster than electron detrapping. The following two basic stages take place during the fixing process.

In the first stage, at temperatures ranging within ≈ 100 °C to ≈ 180 °C, proton migration proceeds till the trapped-electron space-charge field is compensated. On cooling down to room temperature, proton migration is frozen and the pattern of protonic charge, matching the initial trapped electron pattern, is said to be *fixed*. This first stage of the fixing process has been illustrated in Figs. 2a,b.

Recording can also be performed at high temperature (between ≈ 100 °C and ≈ 180 °C) instead of at room temperature. In that case the amplitudes of both trapped electron and proton patterns keep increasing, because they electrically compensate each other during the process, until a final saturating field is reached [21,22]. This results in much higher final amplitudes for both patterns, i. e. higher fixing efficiency, although electrical breakdown can happen under certain circumstances [19].

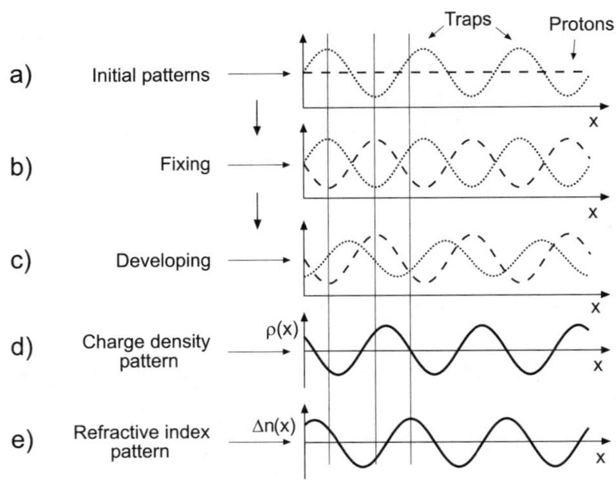

Fig. 2. Stages of photorefractive fixing. (**a**) Initial trap pattern recorded at room temperature, together with a homogeneous proton distribution. (**b**) During thermal fixing ($\lesssim 100\,^\circ\mathrm{C} < T \lesssim 180\,^\circ\mathrm{C}$), ions migrate in the space-charge field until charge compensation is achieved (electrons remain trapped). (**c**) During developing with homogeneous illumination, migration of photoelectrons produces an amplitude decrease and a phase shift of the trap pattern. (**d**) Charge-density pattern resulting from the proton and trap patterns, and (**e**) induced refractive index pattern

In the second stage, the proton pattern is *developed* at room temperature by homogeneous illumination, which gives rise to some redistribution of trapped electrons. Electron redistribution involves both amplitude decrease and phase shift of the electronic grating with respect to the protonic one. The phase shift is due to the photovoltaic effect and becomes particularly important in $LiNbO_3$ [23,24]. The result is a stable index grating originated by the partially uncompensated protonic grating (see a microphotometric investigation in reference [25] and a macroscopic confirmation in [19]). Figures 2c–e schematically illustrates the process. Further homogeneous illumination during later readings produce additional developing and will not at all erase the fixed-developed pattern. When recording is performed at room temperature, the method is called *low-high-low* in reference to the temperature of each step. On the other hand, if recording is performed at the same temperature as fixing, the method is called *high-low*. A novel heating method for fixing via a CO_2 laser beam has been reported [26] as a fast and efficient one for *in situ* local holograms.

In this chapter we will consider only the standard model that accounts for most data in $LiNbO_3$, although very short digressions out of the model will be allowed.

4.2 Other Mechanisms

There are some experimental situations which are not considered by the above standard model. Depending on material conditions, more than one type of electron trap or proton location may be present in the crystal. For example, reduced $LiNbO_3$ presents some electronic contribution to the room temperature dark conductivity [20] involving Nb_{Li}^{4+} defects as electron donors [27]. (Their contribution is higher for more reduced crystals and practically absent for highly oxidized samples.) Under illumination, typically they are not relevant compared with the Fe^{2+} donors which dominate the photoconductivity [28]. These traps are not considered in the ordinary Fe^{2+}–Fe^{3+} scheme used by the model. At high Fe concentrations, $Fe^{2+} \rightarrow Fe^{3+}$ tunneling has been claimed to be the main mechanism for dark decay [29].

When the proton concentration is deliberately decreased down to $\approx 10^{23}\,\mathrm{m}^{-3}$ (dehydrated samples), thermal fixing has been attributed to a different ionic species [19] (probably a residual defect like a lithium self-interstitial [30]). In this case the ionic conductivity shows a higher activation energy (1.4 eV) [31,32]. On the other hand, two or more different proton sites are also inferred from the various peaks of the OH^- infrared absorption band [18,33]. Up to the present, there has been neither experimental data on the possible effects of these sites on the transport properties of protons, nor theoretical analysis of this type of situation. In addition, it has been suggested that, apart from the standard electro-optic contribution of the space-charge field to the index grating, protons themselves could contribute up to 10^{-4}

to the index grating, much in the same way as they strongly modify the substrate index in proton exchanged waveguides [34].

4.3 Erasure of Fixed Gratings

A fixed grating has inevitably a finite lifetime. Even in the dark, there always exists a finite crystal conductivity σ_d which causes the pattern to fade out with a lifetime ranging from weeks to years, depending on the materials and dopants. Both electronic and ionic conductivities follow an Arrhenius-like dependence on temperature T, i. e.

$$\sigma_\mathrm{d} = q n_\mathrm{c} \mu_\mathrm{c} \mathrm{e}^{-\varepsilon_\mathrm{a}/(k_\mathrm{B}T)} \,. \tag{1}$$

In this expression, q is the carrier charge, n_c is the electronic (n) or ionic (H) carrier concentration, μ_c is the electron (μ_{e0}) or ion (μ_{H0}) mobility, ε_a is the activation energy and k_B is the Boltzmann constant. Depending on the values of the parameters appearing in (1), especially the activation energy and the temperature, the dark conductivity will be dominated by the ion or electron contribution. In typical $LiNbO_3$, where ε_a is about 1 eV for protons and about 1.3 eV for electrons, the dark conductivity is mainly protonic up to $\approx$ 180 °C and mainly electronic above this temperature (see below, more details in Sect. 6).

On the other hand, the developed pattern can be delibeately erased by either coherent recording of an appropriately mismatched light pattern or, most often, heating above 200 °C for full electron thermal detrapping and homogeneous electron and proton redistribution. The last treatment gives rise to a completely fresh sample without any memory of the previously stored information.

4.4 Fixing in Other Materials

Photorefractive fixing has been observed in a number of other materials such as $BaTiO_3$ [35,36,37], $(Sr_{0.75}Ba_{0.25})Nb_2O_6$ [38], $Bi_{12}SiO_{20}$ [39,40], $LiTaO_3$ [41], $KNbO_3$ [42,43], $K_{1-y}Li_yTa_{1-x}Nb_xO_3$ [44] and $La_3Ga_5SiO_{14}$:Pr^{3+} [45]. Although in these materials protons are also possible candidates to act as compensating ions in thermal fixing, the actual ions have not been definitely identified as yet.

5 Mathematical Formulation of the Model

A number of authors have investigated the mathematical formulation of the physical model described above for thermal fixing. Early formulations [46,47] explained some aspects of the fixing kinetics. The main limitations of those models were that they did not adequately deal with the photovoltaic transport mechanism and dealt only qualitatively with the developing process, apart from ignoring electron detrapping. Later models are more complete, studying in detail the developing process in a non-photovoltaic material such as SBN [48] and taking into account the photovoltaic effect (which plays a relevant role in the photorefractive effect in $LiNbO_3$) [49,22] as well as electron detrapping [21,22]. (Dark electronic conductivity has been shown to be of key importance to account for photorefractive phenomena in the high temperature range ($\approx 200\,°C$) [50].) Particular, the model developed by *Carrascosa* and *Aguló-López* [21,23] and *Sturman* et al. [22] has provided a plausible explanation of most fixing features.

5.1 General Equations

For the one-dimensional situation, the quoted model [21,22,23] involves the rate and transport equations for free carriers, donors, acceptors and protons, apart from the Poisson and continuity equations, i. e.

$$\frac{\partial n}{\partial t} = (S_{\mathrm{T}} + S_{\mathrm{ph}} I)(N - N^+) - S_{\mathrm{r}} n N^+ - \frac{1}{e}\frac{\partial j_{\mathrm{e}}}{\partial x}\,, \tag{2}$$

$$-\frac{\partial (N - N^+)}{\partial t} = \frac{\partial N^+}{\partial t} = (S_{\mathrm{T}} + S_{\mathrm{ph}} I)(N - N^+) - S_{\mathrm{r}} n N^+\,, \tag{3}$$

$$\frac{\partial E}{\partial x} = \frac{\rho}{\epsilon}\,, \tag{4}$$

$$\frac{\partial j_{\mathrm{H}}}{\partial x} = -e\frac{\partial H}{\partial t}\,, \tag{5}$$

$$j_{\mathrm{e}} = e\mu_{\mathrm{e}} n E + eD_{\mathrm{e}}\frac{\partial n}{\partial x} + eS_{\mathrm{ph}} L_{\mathrm{pv}} (N - N^+) I\,, \tag{6}$$

$$j_{\mathrm{H}} = e\mu_{\mathrm{H}} H E - eD_{\mathrm{H}}\frac{\partial H}{\partial x}\,. \tag{7}$$

In these equations both thermal and optical electron excitation are included. The meanings of the symbols used are as follows:

n, electronic carrier concentration;

$S_{\mathrm{T}} = S_{\mathrm{T0}} \exp[-\varepsilon_{\mathrm{D}}/(k_{\mathrm{B}}T)]$, thermal ionization probability of donors per unit time;

S_{ph}, photoionization cross section;

I, photon flux;

$(N - N^+)$, electron donor concentration;

S_{r}, trapping coefficient for acceptors;

N^+, ionized donor (electron acceptor) concentration;

$n_0 = (S_\mathrm{T} + S_\mathrm{ph} I_0) N_\mathrm{D}/(S_\mathrm{r} N_\mathrm{A})$, average electron concentration;

$N_\mathrm{D} =< N - N^+ >$, average donor concentration;

$N_\mathrm{A} =< N^+ >$, average acceptor concentration;

$N = N_\mathrm{A} + N_\mathrm{D}$, total trap concentration;

e, absolute value of the electron charge;

j_e, electron current density;

E, field induced by charge distribution;

ρ, photoinduced charge density, which includes the contributions of free electrons, active impurities, and protons;

ϵ, dielectric constant of the material;

j_H, proton current density;

H, proton concentration;

H_0, average proton concentration;

μ_e, electron mobility;

$D_\mathrm{e} = D_\mathrm{e0} \exp[-\varepsilon_\mathrm{e}/(k_\mathrm{B}T)]$, effective diffusion constant for electrons;

L_pv, photovoltaic transport length;

μ_H, proton mobility;

$D_\mathrm{H} = D_\mathrm{H0} \exp[-\varepsilon_\mathrm{H}/(k_\mathrm{B}T)]$, effective diffusion constant for protons.

When using light beams of medium and low power, the free electron contribution to the charge density can be neglected and it can be written

$$\rho = (H - H_0 + N^+ - N_\mathrm{A})\mathrm{e} \tag{8}$$

5.2 First-Order Equations: Relaxation Modes

For the practical common case of a spatially periodic one-dimensional light pattern ($\Lambda = 2\pi/K$), it is usual to disregard for all variables harmonics higher than the first one in the Fourier series decomposition. For example, the space-charge field will be written $E(x) = E_k \mathrm{e}^{ikx}$ and so on. This is the linear approximation in the contrast m of the light-fringe pattern. Then, equations (3)–(7) can be cast, after some manipulation, in the form [22]:

$$\frac{\mathrm{d}N_K^+}{\mathrm{d}t} + \overline{\gamma}_\mathrm{e}(1 + \xi_\mathrm{e}) N_K^+ + \overline{\gamma}_\mathrm{e} H_K = F_K\,, \tag{9}$$

$$\frac{\mathrm{d}H_K}{\mathrm{d}t} + \gamma_{\mathrm{H}} N_K^+ + \gamma_{\mathrm{H}}(1+\xi_{\mathrm{H}})H_K = 0\,, \tag{10}$$

where $\overline{\gamma}_{\mathrm{e}} = \gamma_{\mathrm{e}}[1 + K^2 D_{\mathrm{e}}/(S_{\mathrm{r}} N_{\mathrm{A}})]$ (it is very close to γ_{e} as the last term is very small) and the electronic γ_{e} and protonic γ_{H} relaxation rates are given by

$$\gamma_{\mathrm{e}} = \gamma_{\mathrm{e}}^{\mathrm{ph}} + \gamma_{\mathrm{e}}^{\mathrm{T}} = \frac{e\mu_{\mathrm{e}} n_0}{\epsilon}\,, \qquad \gamma_{\mathrm{H}} = \frac{\mathrm{e}\mu_{\mathrm{H}} H_0}{\epsilon}\,. \tag{11}$$

Here, $\gamma_{\mathrm{e}}^{\mathrm{ph}}$ and $\gamma_{\mathrm{e}}^{\mathrm{T}}$ are the photon and thermal contributions to γ_{e}. In addition, the parameters

$$\xi_{\mathrm{e}} = \frac{E_{\mathrm{D}}}{E_{\mathrm{q}}} - \mathrm{i}\frac{E_{\mathrm{pv}}}{E_{\mathrm{q}}}\frac{N_{\mathrm{A}}}{N}\frac{\gamma_{\mathrm{e}}^{\mathrm{ph}}}{\gamma_{\mathrm{e}}}\,, \qquad \xi_{\mathrm{H}} = \frac{E_{\mathrm{D}}}{E_{\mathrm{q}}}\frac{N_{\mathrm{t}}}{H_0} \tag{12}$$

have been also introduced for normalization; in most common cases they are much smaller than one, i. e. $\xi_{\mathrm{e}}, \xi_{\mathrm{H}} \ll 1$. Here, $N_{\mathrm{t}} = N_{\mathrm{A}} N_A D/N$ is the effective trap concentration and

$$E_{\mathrm{D}} = \frac{K k_{\mathrm{B}} T}{e}\,, \qquad E_{\mathrm{q}} = \frac{e N_{\mathrm{t}}}{\epsilon K}\,, \qquad E_{\mathrm{pv}} = \frac{L_{\mathrm{pv}} S_{\mathrm{r}} N_{\mathrm{A}}}{\mu_{\mathrm{e}}} \tag{13}$$

are the so-called diffusion, saturation and photovoltaic fields. The right-hand side of (9) is the effective driving force

$$F_K = -\mathrm{i}\frac{m}{2} N_{\mathrm{t}} \gamma_{\mathrm{e}}^{\mathrm{ph}} \frac{E_{\mathrm{pv}} + \mathrm{i}E_{\mathrm{D}}}{E_q}\,. \tag{14}$$

In order to solve the set of equations, they must be completed with the experimental initial conditions. In principle, they permit us to deal with any design of fixing experiments, including the developing process and the lifetime of the fixed hologram. In the latter case, the set of equations becomes much simpler, since developing is performed at room temperature, where neither proton migration nor thermal electron detrapping occurs. The structure of (9) and (10) indicate that, when $F_K = 0$, the time evolution of the amplitudes N_k, E_k and H_k is characterized by two exponential solutions or *relaxation modes*, proportional to $\mathrm{e}^{-\Gamma_{\mathrm{f,s}}}$, with time constants Γ_{f} (called fast) and Γ_{s} (slow). The values of these time constants depend, apart from the particular experimental conditions, on the stage of the process, i. e. recording, fixing, developing or long-term storage. The useful approximation described below fully exploits the physical meaning of these relaxation modes, and simple expressions are obtained for Γ_{f} and Γ_{s}.

When the pattern modulation is close to one, linearization is not possible. A recent numerical study [51] has concluded that the solution for the first harmonic obtained in the model is representative of the general behavior in most cases. This result is similar to that of a previous study for unfixed gratings [52]. It should also be remarked that the model does not take into account the situation of very long fixing at high temperature, where the modulations of either trap or ion concentrations may reach saturation (no more traps or ions are available for charge compensation).

5.3 A Useful Approximate Solution

In most common cases of fixing, the parameters ξ_e and ξ_H are much smaller than one. Under this condition, the solutions of (9) and (10) are considerably simplified and much physical insight can be gained. In this case, the general expressions for the rate constants are the following:

$$\Gamma_f = \gamma_e + \gamma_H \,, \qquad \Gamma_s = \frac{\gamma_e \gamma_H}{\gamma_e + \gamma_H} (\xi_e + \xi_H) \,. \tag{15}$$

The amplitude of the space-charge electric field E_k will be written in terms of experimental boundary conditions and the time constants Γ_f and Γ_s. For example, for recording starting from a fresh sample (no grating at all),

$$E_k^r(t) = E_f^r(1 - e^{-\Gamma_f^r t}) + E_s^r(1 - e^{-\Gamma_s^r t}) \,, \tag{16}$$

where E_f^r and E_s^r are the components of the stationary amplitude $E_k^r(\infty)$ related to the fast and slow processes. Typically, E_f^r is a quasistationary amplitude, because it is the value reached at the end of the fast stage which afterwards varies very slowly. On erasing, on the other hand, E_k will take the form

$$E_k^e(t) = E_f^e e^{-\Gamma_f^e t} + E_s^e e^{-\Gamma_s^e t} \,, \tag{17}$$

which is applicable to either photon or dark relaxation with appropriate parameters. In the case of photon relaxation, the fast stage corresponds to the developing of the fixed grating.

Expressions (15) also indicate that the fast relaxation constant Γ_f is essentially controlled by the greatest of γ_e^{ph}, γ_e^T and γ_H, whereas the slow relaxation constant Γ_s is basically controlled by the smallest of them.

Although the approximation $\xi_e \ll 1$, $\xi_H \ll 1$ is applicable to most common situations, it is not fully valid for heavily Fe-doped samples in a highly oxidized state, where $\xi_e \lesssim 1$.

6 Experimental Data

A number of fixing experiments have been used to find out important material parameters involved in the model mentioned above, such as the absolute value of the proton concentration H_0 [53] and the dependence of electro-optic coefficients on the wavelength [54]. However, most experimental works have been devoted to investigate different aspects of thermal fixing such as fixing and developing kinetics, the influence of temperature, final diffraction efficiencies and hologram lifetimes. The following sections deal with these aspects in more detail.

6.1 Fixing and Developing Kinetics: Influence of Temperature

According to the model presented above, the fixing process at high temperature is described by a double-exponential time dependence as in (16). This kind of kinetics was observed early on by *Meyer* et al. [46] and *Hertel* et al. [49]. It was found that, within the 176 °C to 180 °C range, a fast initial growth of the diffraction efficiency was followed by a slow increase, according to the theory. As for the kinetics of developing, represented by a single exponential in the model, many studies have reported experimental agreement with the model. An extensive study can be found in a recent work by *De Miguel* and coworkers [24]. In this work, temporal oscillations of the diffraction efficiency were observed before reaching the steady state for high K-vector gratings and strongly oxidized crystals.

On the other hand, expressions (15) indicate that the temperature dependences of the time constants Γ_{f} and Γ_{s} of the exponentials are controlled by the different γ. In turn, in congruent $LiNbO_3$, the temperature dependences of $\gamma_{\mathrm{e}}^{\mathrm{ph}}$, $\gamma_{\mathrm{e}}^{\mathrm{T}}$ and γ_{H} are mainly controlled by the activation energies for electron diffusion ε_{e}, for electron thermal detrapping ε_{D} and for proton diffusion ε_{H} respectively . The relative importance of each γ in determining the values of Γ_{f} and Γ_{s} mainly depends on the relative values of those activation energies. Unfortunately, in $LiNbO_3$ there exists some indetermination in those values.

For example, regarding proton diffusion, even when the experimental conditions allow one to disregard electronic and other ionic contributions to the conductivity, the values of the transport parameters for proton migration are considerably scattered, depending on the authors and methods, as reviewed by *Kovácş* and *Polgár* [55]. Thus, typical values reported for the activation energy ε_{H} range from 0.95 eV in proton-exchanged layers [56] and bulk $LiNbO_3$ [57] to 1.05 eV [58], 1.1 eV [7,59,60], 1.2 eV [16] and 1.23 eV [61] in bulk crystals, and so on. For the diffusion coefficient D_0 the values range from $4\,\mathrm{cm}^2/\mathrm{s}$ [60] to $0.01\,\mathrm{cm}^2/\mathrm{s}$ [56]. A reason for such a wide range of values is the dependence of the activation energy and diffusion coefficient values on the doping and stoichiometry of the samples. The activation energy is reported to decrease [59,55] or to increase [62] with increasing Li/Nb ratio, whereas it increases with increasing Mg concentration [55]. In addition, residual defects like lithium self-interstitials have been proposed to explain the dark conductivity in dehydrated samples with $H_0 < 5 \times 10^{17}\,\mathrm{cm}^{-3}$ [30] with an activation energy of 1.4 eV. Another reason for the value spread is the variety of processes (including electron detrapping) which, because they have similar activation energies, occur in the same temperature range [63]. Therefore, to some extent, the variations in quoted activation energies may reflect the details of the measurement.

On the other hand, regarding electron detrapping very few data are found in the literature. Early measurements of ε_{D} by *Staebler* et al. [64] gave the value of 1.4 eV, whereas a reinterpretation of dark decays measured by *Müller* et al. [65] gave the value of 1.3 eV.

6.2 Diffraction Efficiency of Fixed Gratings

When the developing of a fixed grating is performed without special care, the obtained diffraction efficiency η_{dev} is relatively small, about a factor of one hundred less than the initially recorded value. However, *Staebler* et al. [64,66] early on reported about $\eta_{\text{dev}} \approx 40\%$ with one hologram and about 2% when fixing 500 angle-multiplexed holograms by recording at high temperature. Much later, and recording also at high temperature, efficiencies about 50% were achieved by *Rakuljic* et al. [8] and *Müller* et al. [9] by avoiding shortcircuiting the sample. Finally, *Mendez* and *Arizmendi* [67] have reported a fixing procedure (using recording at high temperature as well) in which they virtually achieved 100% diffraction efficiency with 1 mm thick sample. More recently, the same group has reported a detailed experimental study of the influence of the grating K-vector, the doping level and the reduction state of the impurity on the developing efficiency [24]. The results have been successfully interpreted with the above model, and a high ratio E_{pv}/E_q (photovoltaic field over limiting field) has been found the key factor to produce efficient developing. The role of the grating K-vector is also relevant, as η_{dev} grows substantially as K increases.

6.3 Lifetime of Fixed Holograms

The lifetime of fixed holograms is an issue of major interest for applications. The storage lifetime essentially depends on the ionic conductivity (i. e. proton mobility and proton concentration in $LiNbO_3$) and on the degree of electronic compensation. Again, there is a variety of lifetime values for different samples and different authors, starting from the early lifetime record of 10^5 years reported in [64] where no explanation was given on the procedure. In as-grown $LiNbO_3$, with $H_0 \sim 5 \times 10^{24}\,\text{m}^{-3}$ a storage lifetime of 2 month [32], or, under continuous illumination, 5 years [57] and 15 years [19] have been found. In dehydrated $LiNbO_3$ with $H_0 < 5 \times 10^{23}\,\text{m}^{-3}$ a maximum storage lifetime of 10 years is estimated in [32] and much greater than 15 years in [19].

The value spread can be understood taking into account the high sensitivity of the hologram lifetime on the spatial frequency, doping and reduction state and proton concentration. These dependences are included in the model given in Sect. 4 through the small parameters ξ_{e} and ξ_{H} that appear in the rate constants Γ_{f} and Γ_{s}. Specifically, the hologram lifetime under illumination, τ, is given by [22,57]

$$\frac{1}{\tau} = \Gamma_{\text{s}} = D_{\text{H}} K^2 \left(\frac{H_0}{N_{\text{t}}} + 1 \right) . \tag{18}$$

The above expression indicates that the lifetime is proportional to the square of the grating spacing, $\Lambda^2 = 2\pi/K^2$ and decreases with the ratio H_0/N_{t}. The thermal activation energy of the lifetime is determined by D_{H}, i. e. it is the same as that of proton migration. The K^2 dependence together with

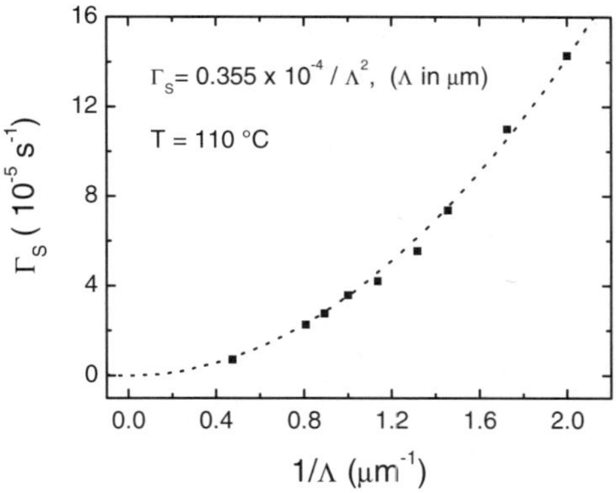

Fig. 3. Experimental dependence of the slow rate constant Γ_s on $1/\Lambda = K/2\pi$ (*solid squares*) and the predicted K^2 dependence (*dashed line*). Taken from [57]

$\varepsilon_H = 0.95\,\text{eV}$ has been experimentally observed in [57] and is illustrated in Fig. 3.

On the other hand, under dark conditions, the developed hologram is compensated by proton migration during the fast process controlled by $\Gamma_f = \gamma_H$ according to (15), where γ_H is given in (11). Thus, the corresponding time constant is independent of K, proportional to $1/H_0$ and with the same activation energy as before, ε_H. This fading of the diffraction efficiency does not indicate that the fixed grating has disappeared only a small fraction of it has been destroyed (approximately that not compensated by trapped electrons). The diffraction efficiency can be recovered by further optical developing as predicted by theory [22] and confirmed by experiments [68]. Complete hologram erasure without any possible retrieval occurs for longer times. This is the slow relaxation process in the dark, which follows a similar expression to (18) but is now controlled by electron detrapping [22].

Finally, it is worth mentioning recent results in Cu-doped Ti:$LiNbO_3$ waveguides [69] (see also [70]), where fixing of gratings at 180 °C with diffraction efficiency above 90% does not require developing at all. In the dark, fixed gratings last unchanged for at least one year and are attributed to migration of Cu ions and not to protons.

7 Optimization of the Fixing Process

An optimized fixed device should simultaneously accomplish two basic objectives: a high diffraction efficiency and a long lifetime of that efficiency either under illumination or in the dark. To this end, judicious criteria must be used to modify the fixing method and/or the crystal parameters. The *high-low* method (recording and fixing at high temperature) appears more appropriate than the *low-high-low* method (recording at room temperature) for producing high diffraction efficiencies. It has, however, the requirement of a delicate stabilization of the experimental setup against thermal turbulences and thermal expansion. As for the lifetime, it has been shown in Sect. 6.3 that the lifetime increases on increasing K^{-2}, and so large grating spacings are advantageous.

Regarding the crystal properties, higher efficiencies can be obtained using heavily doped, highly oxidized samples [24], and long lifetimes are achieved with low proton concentrations (H_0). These two effects are somehow opposite, as the lifetime linearly increases with $(1 + H_0/N_t)^{-1}$. Thus, it increases on decreasing H_0 but decreases on decreasing $N_t \simeq N_D$, which must be small (highly oxidized sample) for high efficiency. Reaching some compromise, lifetimes over 10 (5–10) years have been demonstrated [57]. Well-oxidized samples have the additional advantage of a much smaller electronic dark conductivity [57].

8 Volume Holographic Devices

Photorefractive fixing in $LiNbO_3$ was soon recognized as a very promising technique for long-lifetime volume holographic devices such as for data storage, and, later on, for very narrow-bandwidth interference filters, optical interconnections, optical correlators, etc. More recently, applications of the fixing technique are appearing in the optical waveguide field, with devices analogous to those already shown in bulk crystals, which are useful for telecommunication systems. A number of volume devices are discussed in this section, whereas waveguide devices are discussed in Sect. 9.

8.1 Data Storage

Massive permanent data storage via photorefractive fixing of volume holograms in $LiNbO_3$ was first reported by *Staebler* and co-workers [64,66]. They succeeded in fixing 500 superimposed holograms in an iron-doped $LiNbO_3$ crystal, with low diffraction efficiency and low cross-talk. After more than twenty years, the success in multiplexing more and more non-permanent holograms for photorefractive data storage systems [71,72,73] has renewed the interest in the fixing technique. A digital holographic storage system has been developed [74] incorporating 530 stored and fixed holograms and representing 1.7 MB of digital data. In this case, the postrecording heating procedure (low-high-low) was preferred because it achieves Bragg matching for the entire angular bandwidth of the signal for the perpendicular configuration of the recording beams. Recently, up to 10000 fixed holograms have been multiplexed in $LiNbO_3$ and retrieved with no error by *An* and coworkers [75]. In order to obtain high diffraction efficiencies, a special fixing procedure was used. Instead of heating after all holograms have been stored at room temperature, heating is performed each time a group of ≈ 1000 holograms have been recorded.

8.2 Very Narrow-Bandwidth Interference Filters and Mirrors

A very narrow-bandwidth interference filter was the first photorefractive device entering the market. The device works as a Bragg reflector made of a photorefractive volume grating recorded, fixed and developed in a $LiNbO_3$ crystal. It was independently reported by *Rakuljic* and *Leyva* [8] and *Müller* et al. [9], and commercialized by the group California the through *Accuwave Corporation*. Filters with very narrow spectral bandwidths are currently required in a number of areas: LIDAR, solar and stellar astronomy, high-resolution spectroscopy, extra- or intracavity laser elements and wavelength demultiplexers for optical fiber telecommunication systems. Common interference filters are based on multilayer structures and exhibit bandwidths of a few nanometers with transmission peaks (*bandpass* filters) or reflection peaks (*Bragg reflector* filters) close to 90%. More sophisticated devices, such as Lyot–Ohman and Solc filters, are based on interference with polarized light using birefringent crystals and may reach bandwidths of about 10 pm with transmission peaks of about 50% [76].

The Bragg reflector based on photorefractive volume holography consists of a few millimeter thick Fe:$LiNbO_3$ plate on which a photorefractive grating has been previously fixed. For a typical wavelength of 514.5 nm, the following performances can be achieved: a bandwidth (FWHM) under 10 pm with peak reflectivity above 50% and an angular field of view of about 6°. In Fig. 4, spectral response of the reflectivity of a 50 pm bandwidth filter has been represented. These performances compare well with those exhibited by the polarization interference filters quoted above, and clearly outperform those of dichromate gelatine [77], fiber [78], and waveguide [79] filters which also use volume holography techniques. The photorefractive interference filter shows the additional capabilities of fine thermal or electrical tuning, with tuning coefficients of about 5 pm/ °C and 6 pm/ kV, respectively [80].

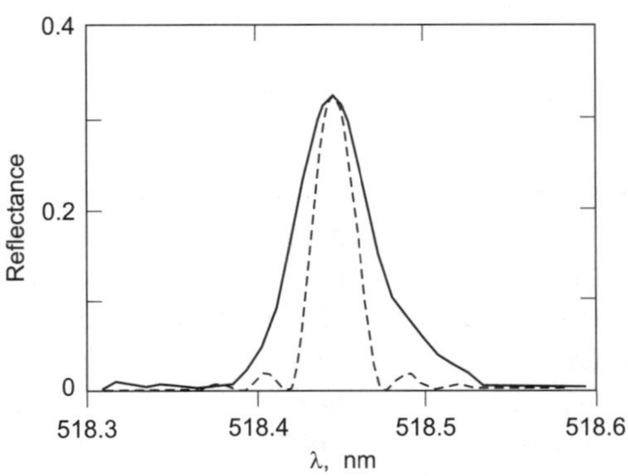

Fig. 4. Spectral response of the reflectivity of a (fixed) photorefractive Bragg reflector: experimental data (*solid line*) and calculated curve (*dashed line*). Taken from [9]

8.3 Wavelength Demultiplexers

Although angle multiplexing for data storage was demonstrated early on at RCA Laboratories [64,66], the potential use of fixing for dense wavelength-division-multiplexing (WDM) has been demonstrated much later by *Breer* et al. [81,82]. They report the demultiplexing of two wavelength channels (channel spacing 0.8 nm) at 1558 nm by Bragg diffraction from two narrow-band fixed reflection holograms in Fe-doped $LiNbO_3$. More than 25 dB cross-talk suppression is reported with intensity losses of a few dB. More recently *Nagel* et al. [83] have designed and implemented a polarization-independent two-channel multiplexer/demultiplexer in a Fe:$LiNbO_3$ crystal. The design consists of four superimposed photorefractive gratings (two per wavelength) in order to satisfy the Bragg condition for both polarization modes of the birefringent $LiNbO_3$. As a proof of concept, preliminary results are reported with diffraction efficiencies of $\approx 50\%$ at 633 nm.

When multiplexing many holograms, the technique of recording and fixing at high temperature (*high-low*) presents clear advantages compared with recording at low temperature (*low-high-low*) [32]. Among these advantages are the following. All holograms show similar diffraction efficiency, i.e. the effective dynamic range of the crystal is much larger, and the photoinduced scattering is almost absent. A possibility only theoretically considered [84] up to now is selective developing and screening of the fixed hologram. Developing should be performed at a much enhanced rate and without appreciable effect on other stored holograms.

8.4 Other Devices

Other volume devices using thermal fixing in $LiNbO_3$ include an all optical holographic switch based on anisotropic diffraction [85] and the first fixed holographic correlator with good performance and high efficiency [86]. The estimated lifetime for the later device is 10 years.

9 Waveguide Holographic Devices

After the development of bulk devices based on photorefractive fixing, a number of papers have appeared demonstrating similar devices based on the guided geometry, which is of more direct interest in telecommunication systems. In a similar way to that described in 8.2, integrated narrow-bandwidth Bragg reflectors have been fabricated by fixing in planar [87] and channel [88] waveguides of Ti:Fe:$LiNbO_3$. A high-low method has been used to fix the photorefractive holograms, obtaining a reflectivity of 60% and a line width of 110 pm. Additionally, the Bragg reflector appears almost insensitive to the light polarization, as illustrated in Fig. 5, where the spectral response of the transmittance has been plotted for TE and TM polarized and unpolarized light. Moreover, thermal tuning has been also demonstrated in these waveguides inside a range of 0.2 nm [89]. Fixed Bragg reflectors have been used as mirrors in an integrated Ti:Er:$LiNbO_3$ laser emitting 5 mW power at 1531.7 nm and pumped with 110 mW at 1480 nm [90].

All these devices can benefit from the fixing technique developed in Cu-doped Ti:$LiNbO_3$ waveguides [69], which apparently gives rise to much longer lifetimes than the conventional technique based on protons. After high-temperature recording with green light, refractive-index changes exceeding $\Delta n \simeq 8 \times 10^{-5}$ for 1550 nm IR light have been achieved in channel waveguides [91,92]. In this case the fixed grating does not need to be developed as the diffraction efficiency after fixing is high it decreases under illumination and increases again in the dark ("dark developing"). Dark developing has also been observed at room temperature in α-phase PE:$LiNbO_3$ waveguides [93], although the lifetime of the grating is rather short in this case. In addition, the last paper gives useful values for most of the photorefractive parameters in PE:$LiNbO_3$ waveguides. These values appear very similar to those of bulk crystals, except for the thermal ionization probability S_T, which is much greater than in bulk crystals. An important aspect related to the space-charge field appearing in the guided configuration is the surface edge effect as theoretically analysed in [94] and experimentally hinted at in [93].

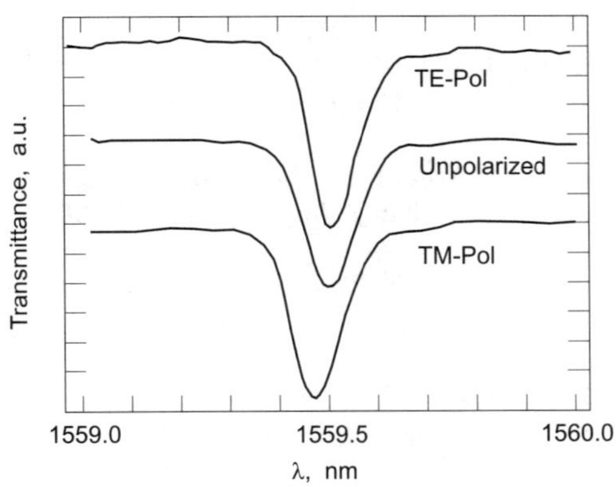

Fig. 5. Transmittance of a fixed photorefractive Bragg reflector versus wavelength for TE, TM and unpolarized light beam. Taken from [88]

10 Summary

This chapter has reviewed our present understanding of thermal fixing phenomena and applications in photorefractive crystals, particularly $LiNbO_3$. A coherent and clear picture has recently emerged and available models can be reliably used to explain and predict experimental results and device performances. A variety of devices, including those needed for telecommunication applications have been proposed and tested, and the way is paved for new developments.

Acknowledgments

This work was supported by the Spanish Ministerio de Educaciòn y Cultura under grants PB98-0061 and PB98-0056, and Comunidad Autònoma de Madrid under grant 07T/0043/2000.

References

1. A. Ashkin, G. D. Boyd, J. M. Dziedzic, R. G. Smith, A. A. Ballmann, H. J. Levinstein, K. Nassau, Appl. Phys. Lett. **9**, 72 (1966)
2. P. Günter, J. P. Huignard: *Photorefractive Materials and Their Applications*, Vol. I and Vol. II (Springer, Berlin, Heidelberg 1988, 1989)
3. F. Agulló-López (Ed.): *Photorefractive Materials*, MRS Bulletin **XIX** (1994)
4. L. Solymar, D. J. Webb, A. Grunnet-Jepsen: *The Physics and Applications of Photorefractive Materials* (Clarendon, Oxford 1996)
5. F. T. S. Yu, F. H. Yin (Eds.): *Photorefractive Optics* (Academic, San Diego 2000)
6. J. J. Amodei, D. L. Staebler, RCA Rev. **33**, 71 (1972)
7. D. L. Staebler, J. J. Amodei, Ferroelectrics **3**, 107 (1972)
8. G. A. Rakuljic, V. Leyva, Opt. Lett. **18**, 459 (1993)
9. R. Müller, M. T. Santos, L. Arizmendi, J. M. Cabrera, J. Phys. D **27**, 241 (1994)
10. D. A. Bryan, R. Gerson, H. E. Tomaschke, Appl. Phys. Lett. **44**, 847 (1984)
11. R. L. Byer, Y. K. Park, R. S. Fergelson, W. L. Kuay, Appl. Phys. Lett. **30**, 17 (1981)
12. M. P. Petrov, S. I. Stepanov, A. V. Khomenko: *Photorefractive Crystals in Coherent Optical Systems*. Springer Ser. Opt. Sci. (Springer, Berlin, Heidelberg 1991)
13. S. I. Stepanov, Rep. Prog. Phys. **57**, 1 (1994)
14. H. J. Coufal, D. Psaltis, G. T. Sincerbox (Eds.): *Holographic Data Storage* (Springer, Berlin, Heidelberg 2000)
15. M. Wessner, C. Herder, D. Kip, Appl. Phys. B **72**, 733 (2001)
16. H. Vormann, G. Weber, S. Kapphan, E. Krätzig, Sol. Stat. Commun. **40**, 543 (1981)
17. N. Schmidt, K. Betzler, M. Grabs, S. Kapphan, F. Klose, J. Appl. Phys. **65**, 1253 (1989)
18. J. M. Cabrera, J. Olivares, M. Carrascosa, J. Rams, R. Müller, E. Diguez, Adv. Phys. **45**, 349 (1996)
19. K. Buse, S. Breer, K. Peithmann, S. Kapphan, M. Gao, E. Krätzig, Phys. Rev. B **56**, 1225 (1997)
20. Y. Ohmori, M. Yamaguchi, K. Yoshina, Y. Inuishi, Jpn. J. Appl. Phys. **15**, 2263 (1976)
21. M. Carrascosa, F. Agulló-López, J. Opt. Soc. Am. B **7**, 2317 (1990)
22. B. I. Sturman, M. Carrascosa, F. Agulló-López, J. Limeres, Phys. Rev. **57**, 12792 (1998)

23. M. Carrascosa, F. Agulló-López, Opt. Commun. **126**, 240 (1996)
24. Eva V. de Miguel, J. Limeres, M. Carrascosa, L. Arizmendi, J. Opt. Soc. Am. B **17**, 1140 (2000)
25. R. Matull, R.A. Rupp, J. Phys. D **21**, 1556 (1988)
26. B. Liu, L. Liu, L. Xu, J. Ma, S. H. Lee, Appl. Opt. **37**, 1342 (1998)
27. O. F. Schirmer, O. Thiemann, M. Wöhlecke, J. Phys. Chem. Solids **52**, 185 (1991)
28. T. Volk, N. Rubinina, M. Wöhlecke, J. Opt. Soc. Am. B **11**, 1681 (1994)
29. I. Nee, M. Müller, K. Buse, E. Krätzig, J. Appl. Phys. **88**, 4282 (2000)
30. D. H. Jundt, M. M. Fejer, R. G. Norwood, P. F. Bordui, J. Appl. Phys. **72**, 3468 (1992)
31. A. Yariv, S. S. Orlov, G. A. Rakuljic, J. Opt. Soc. Am. B **13**, 2513 (1996)
32. S. S. Orlov, W. Phillips: in H. J. Coufal, D. Psaltis, G. T. Sincerbox (Eds.): *Holographic Data Storage* (Springer, Berlin, Heidelberg 2000)
33. L. Kováçs, M. Wöhlecke, A. Jovanovic, K. Polgár, S. Kapphan, J. Phys. Chem. Solids **52**, 797 (1991)
34. S. Kapphan, A. Breithopf, Phys. Stat. Solidi A **133**, 159 (1992)
35. F. Micheron, G. Bismuth, Appl. Phys. Lett. **20**, 79 (1972)
36. D. Kirillov, J. Feinberg, Opt. Lett. **16**, 1520 (1991)
37. D. Zhang, Y. Zhang, C. Li, Y. Chen, Y. Zhu, Appl. Opt. **34**, 5241 (1995)
38. F. Micheron, G. Bismuth, Appl. Phys. Lett. **24**, 71 (1973)
39. L. Arizmendi, R. C. Powell, J. Appl. Phys. **62**, 896 (1987)
40. L. Arizmendi, J. Appl. Phys. **65**, 423 (1989)
41. E. Krätzig, R. Orlowski, Appl. Phys. **15**, 133 (1978)
42. G. Montemezzani, P. Günter, J. Opt. Soc. Am. B **7**, 2323 (1990)
43. G. Montemezzani, M. Zgonik, P. Günter, J. Opt. Soc. Am. B **10**, 171 (1993)
44. V. Leyva, D. Engin, Tong Xiaolin, Tong Min, A. Yariv, A. Agranat, Opt. Lett. **20**, 1319 (1995)
45. T. Nikolajsen, P. M. Johansen, Opt. Lett. **24**, 1419 (1999)
46. W. Meyer, P. Würfel, R. Munser, G. Müller-Vogt, Phys. Stat. Solidi A **53**, 171 (1979)
47. V. V. Kulikov, S. I. Stepanov, Sov. Phys. Solid State **21**, 1849 (1979)
48. M. Jeganathan, L. Hesselink, J. Opt. Soc. Am. B **11**, 1791 (1994)
49. P. Hertel, K. H. Ringhofer, H. Sommerfeldt, Phys. Stat. Solidi A **104**, 855 (1987)
50. M. Carrascosa, L. Arizmendi, J. Appl. Phys. **73**, 2709 (1993)
51. J. Limeres, E. García de la Cera, E. M. de Miguel-Sanz, M. Carrascosa, L. Arizmendi, OSA Trends Opt. Photon. **68**, 320 (2001)
52. E. Serrano, V. Lopez, M. Carrascosa, F. Agulló-López, IEEE J. Quant. Electron. **30**, 875 (1994)
53. R. Müller, L. Arizmendi, M. Carrascosa, J. M. Cabrera, Appl. Phys. Lett. **60**, 3212 (1992)
54. S. Fries, S. Bauschulte, Phys. Satus Solidi A **125**, 369 (1991)
55. L. Kováçs, K. Polgár, in *Properties of Lithium Niobate*, EMIS Datareview Ser. **5**, 109, RN-16037 (1989)
56. C. Canali, A. Carnera, G. Della Mea, P. Mazzoldi, S. M. Al-Shukri, A. C. G. Nut, R. M. De La Rue, J. Appl. Phys. **59**, 2643 (1986)
57. L. Arizmendi, E. M. de Miguel-Sanz, M. Carrascosa, Opt. Lett. **23**, 960 (1998)
58. M. A. Ellabban, G. Mandula, M. Fally, R. A. Rupp, L. Kováçs, Appl. Phys. Lett. **78**, 844 (2001)

59. W. Bollmann, H. J. Stöhr, Phys. Stat. Solidi A **39**, 477 (1977)
60. L. Kováçs, K. Polgár, R. Capelletti, C. Mora, Phys. Stat. Solidi A **120**, 97 (1990)
61. S. Klauer, M. Wöhlecke, S. Kapphan, Phys. Rev. **B45**, 2786 (1992)
62. A. Mehta, E. K Chang, D. M. Smyth, J. Mater. Res. **6**, 851 (1991)
63. L. Arizmendi, P. D. Townsend, M. Carrascosa, J. A. Baquedano, J. M. Cabrera, J. Phys. **3**, 5399 (1991)
64. D. L. Staebler, W. J. Burke, W. Phillips, J. J. Amodei, Appl. Phys. Lett. **26**, 182 (1975)
65. R. Müller, M. Carrascosa, L. Arizmendi, J. M. Cabrera, J. Appl. Phys. **77**, 308 (1995)
66. D.L. Staebler, W.J. Burke, W. Phillips, G.A. Alphonse, Opt. Engin. **17**, 313 (1978)
67. A. Méndez, L. Arizmendi, Opt. Mater. **10**, 55 (1998)
68. L. Arizmendi, A. Méndez, J. V. Alvarez-Bravo, Appl. Phys. Lett. **70**, 571 (1997)
69. J. Hukriede, D. Kip, E. Krätzig, Appl. Phys. B **72**, 749 (2001)
70. R. Sommerfeldt, R. A. Rupp, H. Vormann, E. Krätzig, Phys. Stat. Solidi A **99**, K15 (1987)
71. G. Pauliat, G. Roosen, Int. J. Opt. Comput **2**, 271 (1991)
72. L. Hesselink, M. C. Bashaw, Opt. Quantum Electron. **25**, S611 (1993)
73. J.F. Heanue, M.C. Bashaw, L. Hesselink, Science **265**, 749 (1994)
74. J. F. Heanue, M. C. Bashaw, A. J. Daiber, R. Snyder, L. Hesselink, Opt. Lett. **19**, 1615 (1996)
75. X. An, D. Psaltis, G. W. Burr, Appl. Opt. **38**, 286 (1999)
76. P. Yeh, *Optical Waves in Layered Media* (Wiley, New York 1988)
77. D. E. Sheat, J. S. Leggatt, D. J. McCartney, Electron. Lett. **26**, 42 (1990)
78. G. Meltz, W. W. Morey, W. H. Glenn, Opt. Lett. **14**, 823 (1989)
79. J. Süchtig, Electron. Lett. **24**, 844 (1988)
80. R. Müller, J. V. Alvarez-Bravo, L. Arizmendi, J. M. Cabrera, J. Phys. D **27**, 1628 (1994)
81. S. Breer, H. Vogt, I. Nee, K. Buse, Electron. Lett. **34**, 2419 (1998)
82. S. Breer, K. Buse, Appl. Phys. B **66**, 339 (1998)
83. M. Nagel, A. Siahmakoun, J. Chesnut, OSA Trends Opt. Photon Ser. **62**, 77 (2001)
84. M. Carrascosa, F. Agulló-López, Opt. Commun. **151**, 257 (1998)
85. X.Yan, L. Liu, J. Optics A **1**, 431 (1999)
86. E. M. de Miguel-Sanz, M. Tebaldi, S. Granieri, N. Bolognini, L. Arizmendi, Appl. Phys. B **70**, 379 (2000)
87. J. Hukriede, D. Kip, E. Krätzig, Appl. Phys. B **66**, 333 (1998)
88. J. Hukriede, I. Nee, D. Kip, E. Krätzig, Opt. Lett. **23**, 1405 (1998)
89. J. Hukriede, D. Kip, E. Krätzig, Appl. Phys. B **70**, 73 (2000)
90. Ch. Becker, A. Greiner, Th. Oesselke, A. Pape, W. Sohler, H. Suche, Opt. Lett. **23**, 1194 (1998)
91. J. Hukriede, D. Kip, E. Krätzig, J. Opt. A **2**, 481 (2000)
92. K. Pecthmann, J. Hukriede, K. Buse, E. Krätzig, Phys. Rev. B **61**, 4615 (2000)
93. A. Mendez, A. García-Cabañes, M. Carrascosa, J. M. Cabrera, J. Opt. Soc. Am. B **17**, 1412 (2000)
94. Q. Wang, A. K. Hodari, D. A. Temple, J. Appl. Phys. **82**, 545 (1997)

Holographic Reflection Filters in Photorefractive $LiNbO_3$ Channel Waveguides

Detlef Kip and Jörg Hukriede

Osnabrück University, Physics Department
Barbarastraße 7, 49069 Osnabrück, Germany
{dkip,jhukriede}@uos.de

Abstract. Permanent refractive-index gratings in waveguide devices are of considerable interest for optical communication systems that make use of the high spectral selectivity of holographic filters, e.g. dense wavelength division multiplexing (DWDM) or narrow-bandwidth mirrors for integrated waveguide lasers in $LiNbO_3$. Other possible applications include grating couplers and optical sensors. In this contribution we investigate such holographic wavelength filters in Fe- and Cu-doped $LiNbO_3$ channel waveguides. Permanent refractive-index gratings are generated by thermal fixing of holograms in the waveguides. The samples are fabricated by successive in-diffusion of Ti stripes and thin layers of either Fe or Cu. After high-temperature recording with green light, refractive-index changes up to $\Delta n \approx 10^{-4}$ for infrared light (1.55 µm) are obtained, resulting in a reflection efficiency well above 99% for a 15 mm-long grating. Several gratings for different wavelengths can be superimposed in the same sample, which may enable the fabrication of more complex filters, laser mirrors or optical sensors. By changing the sample temperature the reflection wavelength can be tuned by thermal expansion of the grating, and wavelength filters can be switched on and off by applying moderate voltages using the electro-optic effect. Furthermore, we report on a new thermal fixing mechanism that does not need any additional development by homogeneous light illumination and therefore does not suffer from the non-vanishing dark conductivity of the material.

1 Introduction

Lithium niobate ($LiNbO_3$) crystals are promising substrate materials for applications in integrated optics because of their outstanding nonlinear properties, for instance the electro-optic, acousto-optic and photorefractive effects [1,2,3]. For any integrated circuit, channel waveguides are the basic elements, and high-quality waveguides with low damping constants can be fabricated in $LiNbO_3$ by, for example Ti in-diffusion or proton exchange [4,5,6,7]. By these techniques, devices such as waveguide lasers, directional couplers, mode converters and fast light modulators have been realized [8,9,10,11].

$LiNbO_3$ can be doped with various metal impurities to tailor its photorefractive properties for holographic recording [12,13,14]. The most commonly used dopants are Fe and Cu. $LiNbO_3$:Fe and $LiNbO_3$:Cu crystals are sensitive

P. Boffi, P. Piccinin, M. C. Ubaldi (Eds.): Infrared Holography for Optical Communications, Topics Appl. Phys. **86**, 111–128 (2003)

to light in the visible green or blue wavelength region, but show only small effects in the red or near infrared spectral region. Especially in the 1.3 μm and 1.55 μm wavelength band of optical communications no photorefractive effects have been observed for common continuous wave intensities. Doped $LiNbO_3$ volume crystals are usually produced by adding Fe_2O_3 or CuO to the melt during crystal growth. Nevertheless, it is also possible to increase the impurity concentration by in-diffusion of thin evaporated metal layers at high temperatures [15,16]. Here one can take advantage of the loosely-packed crystal structure of $LiNbO_3$, which results in relatively high diffusion constants of impurities in this material. The technique of diffusion doping is favorable for integrated optics because commercially available un-doped $LiNbO_3$ wafers may be used. A specific region can be locally doped, while leaving the rest of the substrate unchanged.

For application to photorefractive crystals, the long-term stability of the device performance has to be guaranteed. However, phase holograms in these crystals suffer from destructive readout when the material is illuminated with light. In order to avoid erasure, different methods have been developed to make the recorded holograms insensitive to light. Interesting results have been obtained by two-color recording [17,18], i.e. sensitization for red or near-infrared light by pre-illumination with visible blue or green light, and in particular by the technique of thermal fixing [19,20]. Thermal fixing is performed by writing a holographic grating at elevated temperatures of about 180 °C. At these temperatures positively charged ions become mobile and compensate for the generated electronic space-charge field [21]. After the sample has been cooled down, homogeneous illumination with visible light yields a quasi-stabilized fixed hologram.

A rather new application of $LiNbO_3$ waveguides are holographic gratings that can be recorded with light of the photosensitive blue and green spectral region and then be used with infrared light [22,23], where the material is no longer sensitive. Here, channel waveguides for infrared light around 1.55 μm wavelength are of particular interest because of their use in optical communications and measurement systems. In particular, permanent reflection Bragg gratings can help to build dense wavelength division multiplexing (DWDM) systems for optical communications, serve as highly spectral-selective mirrors in integrated distributed Bragg reflection (DBR) lasers in Er-doped $LiNbO_3$, or may be used to build optical sensors for temperature and electric field measurement, to detect gases or complex biological molecules. Such sensors make use of the narrow spectral resonance of the holograms, where the filter reflectivity is already changed for very small effective index changes of the guided light.

In this contribution, we will review the work on holographic reflection filters in Fe- and Cu-doped $LiNbO_3$ channel waveguides. In the following section, the different steps for the preparation of photorefractive waveguides are described. Thereafter, techniques and experimental setups for holographic recording and readout of the gratings are explained. The last part presents typical experimental results for the photorefractive properties and performance of holographic reflection gratings in $LiNbO_3$ channel waveguides. This includes the charge-transport properties of diffusion-doped $LiNbO_3$, fundamental filter properties, the possibilities for hologram multiplexing, wavelength tuning and electrical switching of the filters, and a discussion of the long-term stability of thermally-fixed gratings.

2 Sample Preparation

The fabrication of photorefractive channel waveguides in $LiNbO_3$ crystals consists of several steps that have to be performed in a clean environment. As substrate material, pieces of commercially available x- or y-cut $LiNbO_3$ wafers can be used. Typical dimensions are $7 \times 17\,\mathrm{mm}^2$, which, after the necessary polishing of the endfaces, results in samples with a length of about 15 mm. The ferroelectric c-axis coincides with the propagation direction of the light and points along the longer side. Single-mode waveguides are fabricated by Ti in-diffusion. At first a thin Ti layer with a typical thickness of 100 nm is deposited onto the polished surface using electron-beam evaporation. Then lithographic techniques are used to pattern the Ti film into narrow stripes with a width of 6 to 8 µm. For larger stripe widths or thicker evaporated Ti films, higher modes of the waveguide can appear, and in this case more than one filter resonance will be observed. Typical parameters for Ti in-diffusion is an annealing time of 22 h at a temperature of 1000 °C in air. In this way single-mode waveguides for 1.55 µm wavelength with losses as low as 0.08 dB/ cm are formed. They guide one TE and one TM mode, which are both ordinarily polarized. An optimization of the waveguide parameters results in almost equal propagation constants of the two guided modes. This is of great importance for the filter performance because in this case the peak wavelength of the filter does not depend on the input polarization. Furthermore, the size of the light-field distribution of the single-mode channel waveguides matches well with the typical diameter of non-polarizing single-mode fibers for 1.55 µm, and a field overlap for direct fiber-chip coupling higher then 0.9 can be obtained in this way.

Doping of the surface layer is done by in-diffusion of thin films of either metallic Fe or Cu [15,16]. Layers of 5 to 100 nm thickness are evaporated onto the sample surface and in-diffused for several hours at temperatures of 1000 °C. The in-diffusion process can be either performed in an Ar atmosphere to reduce the crystals, i.e. to get a larger ratio of Fe^{2+}/Fe^{3+} and Cu^{+}/Cu^{2+}, respectively, or it can be performed in an O_2 atmosphere to get a smaller ratio. The diffusion constants of Fe and Cu into $LiNbO_3$ at a temperature of 1000 °C are $D_{\mathrm{Fe}} = (1.8 \pm 0.2) \times 10^{-3}\,\mu\mathrm{m}^2\mathrm{s}^{-1}$ and $D_{\mathrm{Cu}} = (1.0 \pm 0.2)\,\mu\mathrm{m}^2\mathrm{s}^{-1}$ [15,16]. Both are much larger then the value for Ti, which is only $D_{Ti} = (4.4 \pm 0.5) \times 10^{-5}\,\mu\mathrm{m}^2\mathrm{s}^{-1}$ [15]. As a consequence, Cu doping of the sample surface has to be performed after the channel waveguides are fabricated by Ti in-diffusion, whereas Fe doping can be done also before this step. In both cases, a nearly constant doping profile of either Fe or Cu is obtained in the range of the waveguide depth. This depth is defined by the Ti diffusion profile.

Typical doping concentrations in the waveguiding layer are in the range of 1 to 6×10^{25} atoms/ m^3 for Fe and 0.4 to 6×10^{25} atoms/m^3 for Cu, respectively [22,24]. However, there exists a practical threshold for the amount of Cu that can be in-diffused into the surface of the substrate. For thicker layers of the evaporated Cu, diffusion leads to a decrease of surface quality and an increased Cu^{2+} absorption at 1.55 µm. As a result, the damping coefficient of the infrared light that is guided in the channel waveguides can be significantly increased, which may decrease the performance of the component or even prevent its applicability. As an example, in Fig. 1 surface scans of two different Cu-doped samples, measured with an atomic force microscope (AFM), are shown [24]. For lower doping ($c_{\mathrm{Cu}} = 1.1\times10^{25}$ atoms/ m^3) the surface is still smooth and almost comparable to that of a sample without additional doping. When the doping is further increased ($c_{\mathrm{Cu}} = 4.7\times10^{25}$ atoms/ m^3), the surface roughness strongly increases. For samples with such a high Cu doping, the damping for 1.55 µm reaches values of 2 dB cm^{-1}. On the other hand, no degradation of the surface or increase of absorption in the infrared is observed for Fe in-diffusion.

Finally, the two endfaces of the waveguide sample have to be polished to enable direct light coupling with optical fibers. After this step the typical length d of the waveguide is about $d = 15$ mm. To increase the coupling efficiency into the waveguide and to reduce multiple reflections of the recording light inside the sample, antireflection coatings for the two endfaces and for the bottom face can be used. Although the refractive index of MgF_2 with $n_F = 1.34$ (for $\lambda = 1.55$ µm) does not match perfectly to the desired index of

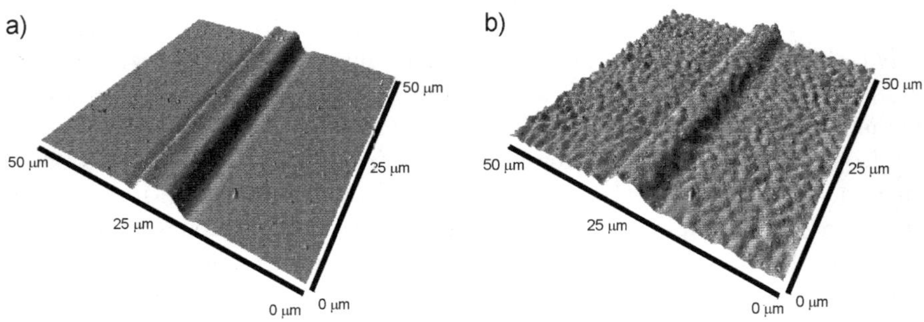

Fig. 1. Surface profiles of $LiNbO_3$:Ti:Cu channel waveguides measured with an AFM. Shown is a part of the 8 µm-wide channel for a sample with (**a**) $c_{\mathrm{Cu}} = 1.1\times10^{25}$ atoms/ m^3 and (**b**) $c_{\mathrm{Cu}} = 4.7\times10^{25}$ atoms/ m^3. The increase in height is due to the Ti in-diffusion into the sample and has an amplitude of about 200 nm

an ideal $\lambda/4$-coating ($n'_F \approx 1.49$), this material produces very resistant layers which adhere well on $LiNbO_3$ without additional heating. Such an annealing treatment has to be avoided in most cases, because otherwise it changes the preset oxidation state of the in-diffused dopant (Fe or Cu). A 290 nm-thick layer of MgF_2 is used as an anti-reflection coating of the endfaces for infrared light. In the same way, a 110 nm-thick layer can be used to reduce the surface reflections of the green light during holographic recording.

3 Holographic Recording and Readout

In this section, we briefly describe the experimental setups for holographic recording and readout of the gratings, as well as the techniques for thermal fixing and development.

In principle, elementary refractive-index gratings can be formed by two guided beams counter-propagating in the waveguide channel. However, as our samples are not sensitive to infrared light, the gratings have to be recorded in the waveguide volume by two external beams of shorter wavelength that impinge upon the surface of the sample. This geometry is of advantage here, because it enables us to adjust the grating period of the hologram by precisely choosing the correct intersection angle of the recording beams. A schematic of the recording and readout geometry is shown in Fig. 2. Two expanded light beams of the green line ($\lambda_w = 514.5\,\text{nm}$) of an Ar ion laser interfere inside the sample. The grating vector $K = 2\pi/\Lambda$ is directed along the c-axis and the recording light is ordinarily polarized to reduce holographic scattering and beam fanning. For the grating period we get

$$\Lambda = \lambda_w/(2\sin\Theta)\,, \tag{1}$$

where 2Θ is the (full) intersection angle of the two beams outside the sample. With an angle of $2\Theta \approx 90°$, readout of the gratings in a linear reflection geometry is possible using (guided) infrared light with a wavelength around 1.55 μm.

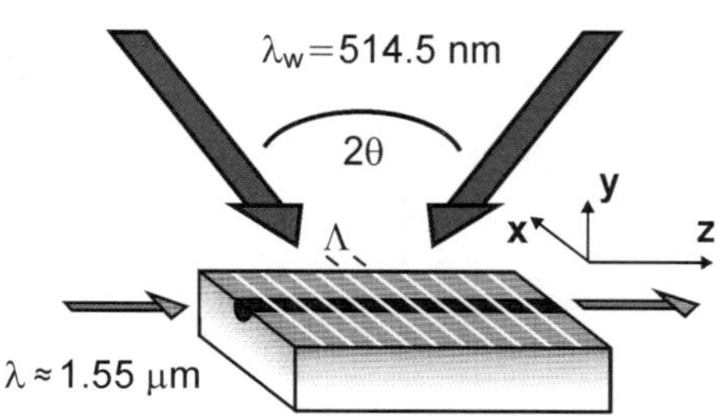

Fig. 2. Schematic of the recording and readout geometry for the holographic measurements. The gratings recorded with green light form reflection gratings for the infrared readout light in the channel waveguides

Thermal fixing is performed by heating the waveguide sample to 180 °C while the grating is written [19,20]. In this way, positively charged ions (e.g., protons) become mobile and can compensate for the generated electronic space-charge field [21]. Because of the relatively long recording times of up to 2 h an active phase stabilization has to be used [22,26,27]. This system compensates for thermal drift of the sample holder and other optical components and thus ensures a stable light pattern during recording.

Special care has to be taken to precisely adjust the angle of 2Θ during recording, because this defines the position of the peak wavelength λ_p for readout of the grating:

$$\lambda_\mathrm{p} = 2n_\mathrm{eff}(T, c_\mathrm{Fe/Cu})\Lambda(T, \Theta)\,. \tag{2}$$

Although the effective refractive index n_eff of the guided light is predominantly defined by the diffusion profile of Ti, which is constant for all samples investigated here, both Fe and Cu doping results in a slightly shifted value of n_eff [15,24]. Next, the thermal expansion of $LiNbO_3$ at higher temperatures T must be taken into account, too, because this directly changes the grating period Λ [23]. We can roughly estimate that the peak wavelength λ_p of a grating recorded at 180 °C is reduced by 1 nm when the sample is cooled down to room temperature. Last, the refractive indices of $LiNbO_3$ depend on temperature, but for the ordinary index that is used here this influence can be neglected when compared with the effect of thermal expansion.

After recording, the samples are cooled down to room temperature within a few minutes. The fixed gratings can be developed by homogeneous illumination with the expanded light beam of an Ar ion laser. Alternatively, the incoherent light of a halogen lamp can be used. The holograms are read in reflection geometry in the waveguide channels with the guided, ordinarily polarized TE and/or TM mode. In the following we have used a Distributed Feedback (DFB) laser which can be tuned over a small wavelength range around 1557 nm. At the peak wavelength λ_p the Bragg condition for readout, $\Lambda = \lambda_\mathrm{p}/2n_\mathrm{eff}$, is exactly fulfilled, and the transmission drops down from unity to a minimum at $T(\lambda_\mathrm{p}) = 1 - \eta$. The diffraction efficiency η at the peak wavelength λ_p is given by the Kogelnik equation [25]

$$\eta(\lambda_\mathrm{p}) = \tanh^2(\pi \Delta\, n\; d/\lambda_p)\,. \tag{3}$$

Here $\Delta\, n$ is the amplitude of the refractive-index change and d is the grating length.

4 Experimental Results

In this section the experimental results of sample characterization will be summarized. At first we present some fundamental properties of photorefractive Ti in-diffused waveguides in $LiNbO_3$. Then the recorded holographic filters will be analyzed and the possibilities for hologram multiplexing, thermal tuning of peak wavelengths and electrical switching of the filter will be discussed. The last part describes the influence of dark compensation on the grating relaxation, i.e. the long-term stability of the gratings. Moreover, a new method for thermal fixing that does not need any further development process will be introduced.

4.1 Photorefractive Properties

For optimization of the filter properties a detailed knowledge of the charge transport mechanism and fundamental photorefractive properties of the waveguide samples is necessary. Here, results from volume-doped crystals cannot be simply transferred because of the high Ti concentration in the waveguiding layer and, especially for some of the samples, the high doping level with either Fe or Cu. Furthermore, the doping process by in-diffusion of metal layers instead of doping during growth of the crystals may result in different impurity sites in the crystal lattice, which can influence the charge transport properties too.

Light-induced refractive-index changes can be measured by recording the holographic gratings at room temperature. If gratings are written for different recording times t, the saturated refractive-index change Δn_s can be obtained by fitting an exponential function of the form

$$\Delta n(t) = \Delta n_s[1 - \exp(t/\tau)] \tag{4}$$

to the measured data $\Delta n(t)$. An example for different Cu-doped samples is given in Fig. 3a. When plotting the saturated refractive-index change Δn_s as a function of the Cu concentration c_{Cu} of different samples in Fig. 3b, a linear dependence is found. We can assume an almost constant ratio $c_{Cu^+}/c_{Cu^{2+}}$ for all the investigated waveguides, because all the samples have been thermally treated in the same way during fabrication. This assumption leads to a linear dependence $\Delta n_s \propto c_{Cu^{2+}}$, which is in agreement with a one-center model of charge transport [28].

The photoconductivity of the photorefractive waveguides can be determined by measuring the decay of the gratings during homogeneous light illumination. First, holographic gratings are written at room temperature with green light, and the diffraction efficiency η and the peak wavelength λ_p for for example, TM-polarized infrared light, for example, are measured. Then the sample is illuminated homogeneously off-Bragg with green light, and in time intervals of a few seconds the diffraction efficiency at the peak wavelength is determined again. The temporal development of the refractive-index change $\Delta n(t)$ is obtained using the Kogelnik equation in (3) (see Fig. 4c). From this dependence we can calculate the photoconductivity σ_{ph} using the relations

$$\Delta n(t) = \Delta n^0 \exp(-t/\tau) \,, \tag{5}$$

$$\sigma_{ph} = \epsilon\epsilon_0/\tau \,. \tag{6}$$

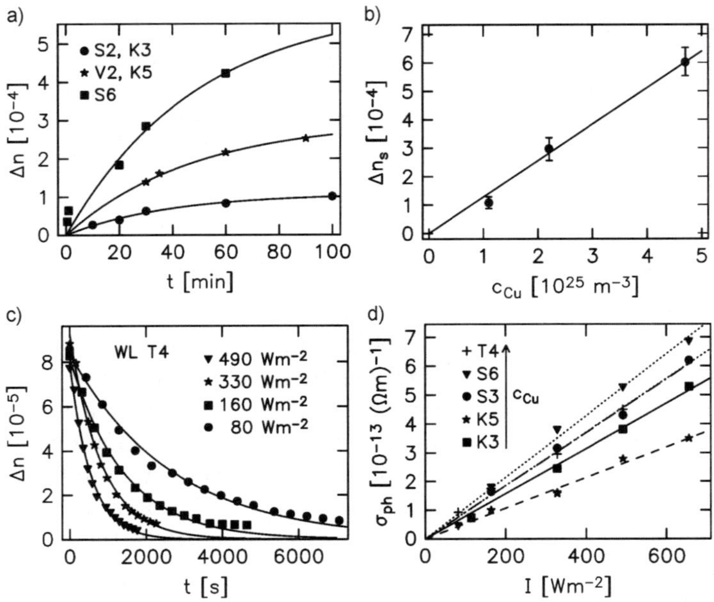

Fig. 3. Light-induced refractive-index changes and photoconductivity in Cu-doped $LiNbO_3$ waveguides at room temperature: (**a**) Refractive-index change Δn versus recording time t for different samples (S2/K3: $c_{Cu} = 1.1 \times 10^{25}\,m^{-3}$; V2/K5: $c_{Cu} = 2.2 \times 10^{25}\,m^{-3}$; S6: $c_{Cu} = 4.7 \times 10^{25}\,m^{-3}$), (**b**) saturated refractive-index change Δn_s versus copper concentration c_{Cu}, (**c**) decay of refractive-index gratings Δn because of homogeneous illumination with green light ($\lambda_w = 514.5\,nm$) of varying intensity for the sample T4 ($c_{Cu} = 5.7 \times 10^{25}\,m^{-3}$), and (**d**) photoconductivity σ_{ph} as a function of erasure intensity I for different Cu-doped samples (S3: $c_{Cu} = 3.2 \times 10^{25}\,m^{-3}$

Here Δn^0 is the refractive-index change for $t = 0$, ϵ_0 is the permittivity of free space, and $\epsilon = 28$ is the appropriate dielectric constant of $LiNbO_3$. In Fig. 4(d) the photoconductivity σ_h is given as a function of the erasure intensity I ($\lambda = 514.5\,nm$) for different Cu-doped samples. As can be seen, the photoconductivity depends linearly on intensity, $\sigma_{ph} \propto I$, which is again in good agreement with a one-center model for charge transport. The absolute value of the specific photoconductivity σ^0_{ph}, defined by $\sigma^0_{ph} = \sigma_{ph}/I$, is of the order of some $10^{-16}\,mV^{-2}$, which is comparable to data obtained for homogeneously copper-doped volume samples [29]. Fe-doped $LiNbO_3$ waveguides have been investigated as well, and for this dopant a good agreement of the photorefractive properties with the predictions of a one-center model has been obtained, too.

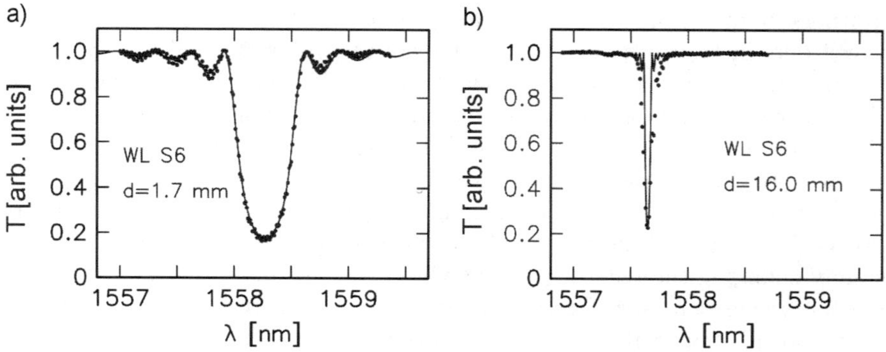

Fig. 4. Transmission T versus wavelength λ for gratings recorded at room temperature in $LiNbO_3$:Ti:Cu channel waveguides: (**a**) grating length $d = 1.7$ mm, recording time 60 min; (**b**) grating length $d = 16.0$ mm, recording time 1 min. The *solid lines* are fits of the wavelength-dependent Kogelnik equation [25]

4.2 Fundamental Filter Properties

Next we will discuss some principal properties of holographic reflection gratings in channel waveguides. The following examples have been obtained in Cu-doped samples, where gratings were recorded with green light and a total writing intensity of 1200 Wm^{-2}. Recording has been performed at room temperature, so the gratings are not thermally fixed. The length d of a grating can be simply adjusted by using an additional aperture (slit) for the recording light placed directly on top of the surface of the waveguide. In Fig. 4a,b the filter characteristics for a short ($d = 1.7$ mm, writing time 60 min) and for a long grating ($d = 16$ mm, writing time 1 min), respectively, are presented. For better comparison, the recording times have been adjusted to get almost equal diffraction efficiencies of about 80% at the respective peak wavelength. As expected, the full-width-half-minimum (FWHM) is strongly decreased for longer gratings and can reach values as low as 0.05 nm. The transmission curves $T(\lambda)$ ("rocking curves") are also in fairly good agreement with the predictions of the wavelength-dependent Kogelnik equation in [25], which are plotted as solid lines in the two graphs with Δn as a free parameter.

Gratings can be read with both, TE- or TM-polarized infrared light. For our standard channel waveguide parameters (6 μm-wide channel, diffusion of a 100 nm-thick Ti layer for 22 h at 1000 °C) and Fe doping we get almost identical effective refractive indices for the the two modes. In this way, a nearly polarization-independent filter can be realized. Here the difference of the corresponding peak wavelengths for the two polarizations is of the order of 0.05 nm. It can be completely eliminated by applying a small bias voltage to the sample and tuning the effective refractive indices via the electro-optic effect.

The filter characteristic can be further tailored, for instance to get nearly rectangular transmission curves $T(\lambda)$, by using both chirped and apodized gratings. Simple chirped gratings can be fabricated by using slightly curved wavefronts instead of plane waves for recording. More complicated gratings can be realized with the help of specially-designed phase masks through which the samples are illuminated during recording with a single laser beam [10]. At the same time, the use of phase masks will result in a recording geometry that is less sensitive to phase fluctuations and instabilities, and thus may be a good alternative to the two-beam holographic setup that has been described in the proceeding section. However, the freedom to adjust the grating period(s) by simply changing the recording angle will be lost in this way. Apodization of the holograms can be realized in different ways. The grating can be written with a locally-varying intensity distribution, which then results in a locally-varying refractive-index change, too. Another possibility is to use a non-homogeneous illumination for the development of the gratings.

4.3 Hologram Multiplexing

In photorefractive crystals several holograms can be superimposed in the same volume, and they can be addressed either by changing the angle under which the readout beam enters the sample or by tuning the readout wavelength. The latter option has to be used for holographic gratings in a linear reflection geometry, i.e. for channel waveguides. Here several gratings with different grating periods can be recorded and thermally fixed, thus reflecting several wavelengths at the same time.

To demonstrate the multiplexing capabilities of photorefractive channel waveguides in $LiNbO_3$, three reflection holograms with a peak spacing of 0.8 nm, corresponding to a 100 GHz channel spacing in DWDM, were superimposed in a Cu-doped sample. The resulting transmission spectrum for TM polarization is plotted in Fig. 5. The holograms were recorded at room temperature one after the other. To compensate for the erasure of the first and second hologram during recording of the latter ones, the respective recording

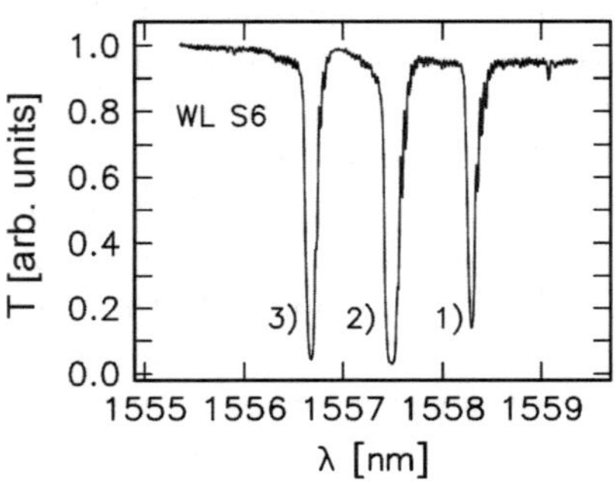

Fig. 5. Three superimposed reflection gratings for infrared light in a $LiNbO_3$:Ti:Cu channel waveguide is the normalized transmission T versus readout wavelength λ. The hologram spacing is 0.8 nm with an average diffraction efficiency of about 93%

times have to be properly adjusted. In the example here they decrease from 16 min for the first hologram (1) to 4 min and 1.5 min for the second (2) and third (3) one, respectively, resulting in an average diffraction efficiency of about 93% for each hologram. The length of the gratings was approximately 16 mm.

4.4 Wavelength Tuning and Electrical Switching

The position of the peak wavelength λ_p of a holographic filter depends on both the grating period Λ and the effective refractive index n_{eff} of the guided light. Therefore, the filter properties can be tuned by heating the sample, i.e. by thermal expansion, or by applying external electric fields, i.e., using the electro-optic effect:

$$\lambda_p(T_s) = \lambda_p(T_0)[1 + \alpha_{33}(T_s - T_0)] , \tag{7}$$

$$\lambda_p(E) = \lambda_p(E = 0)(1 - 0.5\, n_{\text{eff}}^2 r_{22} E) , \tag{8}$$

where $(T_s - T_0)$ is the difference of the sample and reference temperatures, α_{33} is the thermal expansion coefficient along the c-axis, E is the applied electric field along the y-axis, and $r_{22} = -r_{12}$ is the appropriate electro-optic tensor element. While thermal effects, due to their relatively large response time, are well suited for precise tuning of the peak wavelength, the fast response of the electro-optic effect can be used for switching on and off the filter.

Figure 6 presents a measurement of the peak wavelength λ_p of a fixed grating in a $LiNbO_3$:Ti:Fe channel waveguide as a function of the sample temperature T_s. The minimum of the transmitted spectrum $T(\lambda)$ for TM polarization has been evaluated. Obviously, the data follow a straight line, and from a linear fit the thermal expansion coefficient $\alpha_{33} = (4.5 \pm 0.5) \times 10^{-6}\text{K}^{-1}$ can be determined. Here we have neglected the temperature dependence of the effective refractive index n_{eff} for the infrared light in the investigated temperature range.

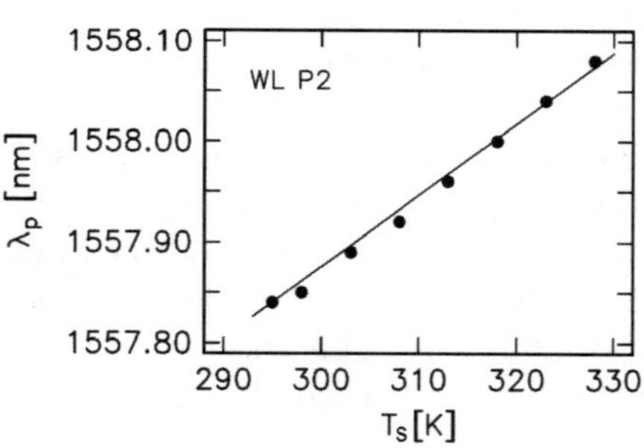

Fig. 6. Peak wavelength λ_p versus sample temperature T_s of a fixed grating in a $LiNbO_3$:Ti:Fe channel waveguide. A linear fit (*solid line*) yields the thermal expansion coefficient $\alpha_{33} = (4.5 \pm 0.5) \times 10^{-6}\text{K}^{-1}$

4.5 Long-Term Stability of Fixed Gratings

The lifetime of thermally-fixed refractive-index gratings is an important factor for the application of integrated holographic filters. A drawback of these devices, when compared with photorefractive volume samples, are the high concentration of Ti in the waveguiding layer, as well as the high doping level of either Fe or Cu that is required to obtain diffraction efficiencies close to 100%. High impurity concentrations lead to a significant increase of the dark conductivity, which in turn causes a relaxation of the thermally fixed and developed holographic gratings.

For samples with no doping or very low doping level, the dark conductivity is mainly caused by the proton conductivity. However, for higher Fe- and Cu-doped $LiNbO_3$ bulk samples and waveguides, the dark conductivity is dominated by the influence of impurities and increases exponentially with doping concentration. Thermally excited electrons can drift in the ionic space-charge field, which finally leads to an almost complete compensation of this field with no net refractive-index change.

In most of the Fe- and some of the Cu-doped samples such a relatively fast dark-relaxation of the fixed gratings is observed. As an example, this effect is shown in Fig. 7 for a grating in a $LiNbO_3$:Ti:Fe channel waveguide. This sample has a dark conductivity of $\sigma_\mathrm{d} \approx 2.0 \times 10^{-16}\,\Omega^{-1}\,\mathrm{m}^{-1}$, which causes a relaxation of the refractive-index change with a time constant of about 50 h. On the other hand, the grating can be redeveloped by homogeneous illumination. Despite the option of a complete redevelopment, the refractive-index change can be adjusted to any desired value $\Delta n' < \Delta n_\mathrm{s}$ by permanent illumination of the sample with light of low intensity, e.g. using blue LEDs. With such illumination, at a certain level of relaxation that is due to the dark conductivity, a balance of dark relaxation and grating development can be reached, i.e. the electronic drift current is compensated by a photovoltaic current generated by the illumination. This effect is shown in Fig. 7 for different intensities of the blue light ($\lambda = 494 \pm 10\,\mathrm{nm}$) filtered out of the spectrum of a halogen lamp with an interference filter. Using the same mechanism but with non-homogeneous illumination of the sample, it is possible to fabricate apodized gratings a with spatially-varying amplitude of the refractive-index change.

An interesting observation that is important for applications has been made during the investigation of reduced Cu-doped waveguides [31]. In these samples, thermally-fixed gratings have been recorded that do not need any development process and that are stable in the dark, i.e. they do not suffer from relaxation because of the high dark conductivity. This effect will be discussed in detail in the following section. In the first experiment, a reflection grating was recorded for 60 min in the oxidized waveguide W3 at a temperature of 180 °C. Thereafter, when the sample had been cooled down to room temperature, the corresponding transmission spectrum $T(\lambda)$ was taken in the infrared. Figure 8a shows this spectrum at the initial stage directly

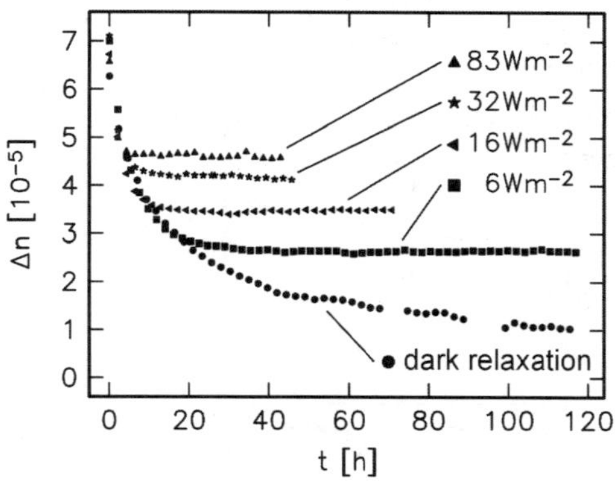

Fig. 7. Compensation of a fixed grating in $LiNbO_3$:Ti:Fe in the dark and under illumination with incoherent blue light ($\lambda = 494 \pm 10\,\text{nm}$). Shown is the temporal evolution of the refractive-index change Δn for different intensities and in the dark. After each measurement the hologram has been redeveloped up to saturation using intensive green light

after recording. In Fig. 8b the spectrum after the grating has been developed by homogeneous illumination with intensive green light is displayed. Without development, only a very small Bragg peak can be seen, which is in good agreement with the behavior expected from the common theory of thermal fixing in $LiNbO_3$ [19,20,21]. When the fixed hologram is developed until saturation by homogeneous illumination, the reflection efficiency increases dramatically. From the measured diffraction efficiency of $\eta = 98.5\%$ for the infrared light we can calculate a refractive-index change of $\Delta n = 1.2 \times 10^{-4}$ in the waveguiding channel by using (3).

The same measurement has been repeated with the reduced waveguide T4. In this sample, the copper concentration is more than two times higher than in W3, and the ratio $c_{Cu^+}/c_{Cu^{2+}}$ is much larger. A grating was recorded for 120 min at 180 °C, and again, directly after cooling down, the transmission spectrum was examined. As can be seen in Fig. 8c, a strong peak was identified and the measured diffraction efficiency reached 95.6%, corresponding to a refractive-index change of about $\Delta n = 8 \times 10^{-5}$. It should be emphasized that at this stage no development of the grating had performed. The fixed grating in this sample was then is additionally developed with green light. The measured spectrum is given in Fig. 8d, where the maximum reflection efficiency is now considerably decreased to a value of 50.2%.

The corresponding temporal behavior of the refractive-index change $\Delta n(t)$ during development is non-monotonous: the value of Δn first drops rapidly, then passes through a minimum and finally reaches a steady-state value that is still much smaller than the initial value Δn^0. The completely developed refractive-index grating of the reduced sample T4 was kept at room temperature in the dark and the temporal evolution of Δn was again moni-

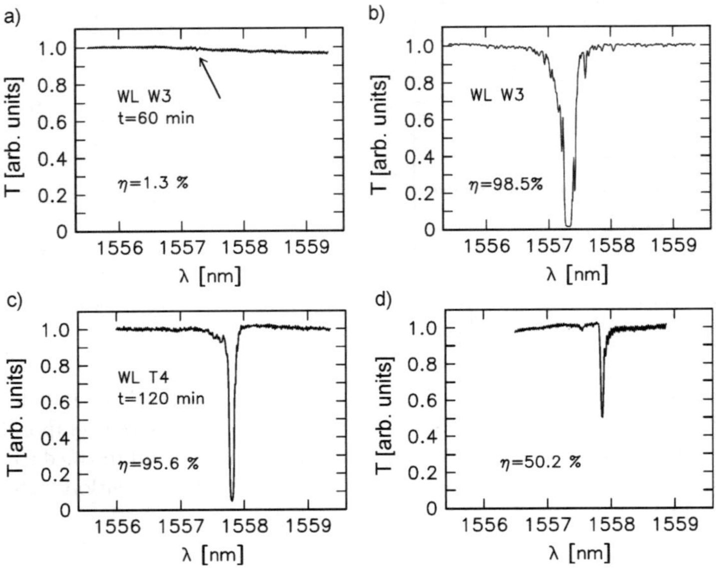

Fig. 8. Transmission T before [(**a**) and (**c**)] and after [(**b**) and (**d**)] the development process versus the readout wavelength λ for gratings recorded in the Cu-doped, oxidized sample W3 (upper plots (**a**) and (**b**); recording time 60 min) and in the Cu-doped, reduced sample T4 (lower plots (**c**) and (**d**); recording time 120 min). For the oxidized sample, the development leads to an increase of diffraction efficiency from 1.3% to 98.5%, whereas for the reduced sample a decrease of diffraction efficiency from 95.6% to 50.2% is measured after development. The oxidized sample with $c_{\mathrm{Cu}} = 2.2 \times 10^{25}\,\mathrm{m}^{-3}$ was annealed in dry O_2 atmosphere, while the reduced one with $c_{\mathrm{Cu}} = 5.7 \times 10^{25}\mathrm{m}^{-3}$ was annealed in a wet Ar atmosphere

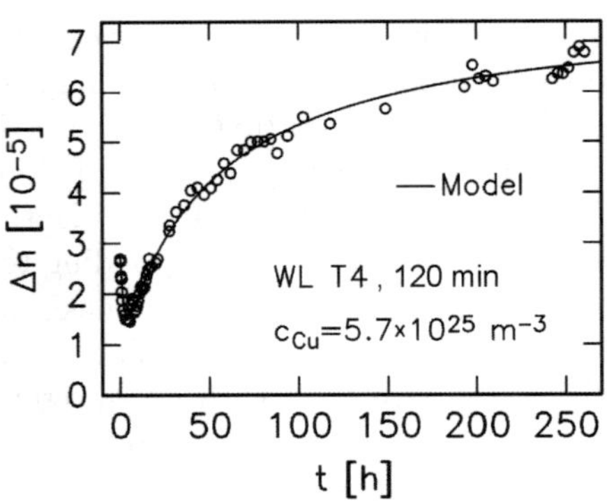

Fig. 9. Dark compensation of a fixed and developed grating in a reduced $LiNbO_3$:Ti:Cu channel waveguide. Shown is the temporal development of the refractive-index change Δn. The *solid line* is a fit according to (10)

tored over a total range of 250 h. The result is given in Fig. 9 and shows the reverse process: Δn first decreases, passes through a minimum and grows again before it finally saturates. Now the saturation value is larger than the start value. In the dark Δn evolves back to its initial value Δn^0 before the development process. In the dark (without development) the initial refractive-index changes Δn^0 was stable, and no decrease was measured over a time of more than one year.

For an explanation of the observed temporal evolution of Δn we assume a superposition of two different refractive-index gratings Δn_b and Δn_c that are out of phase by $(180° - \phi)$. The corresponding model is illustrated in Fig. 10. The grating Δn_c grows already during high temperature recording and remains constant after cooling down. At room temperature it is not affected by light anymore. The second grating Δn_b is generated during the development process. It grows exponentially until saturation with a time constant τ_1 while the sample is illuminated homogeneously. The time evolution of the refractive-index change is thus described by

$$n(t) = \left(\left\{ \Delta n_\mathrm{c} - \Delta n_\mathrm{b}^0 \cos\phi \left[1 - \exp(-t/\tau_1)\right] \right\}^2 + \left\{ \Delta n_\mathrm{b}^0 \sin\phi \left[1 - \exp(-t/\tau_1)\right] \right\}^2 \right)^{-\frac{1}{2}} . \tag{9}$$

Correspondingly, the time evolution of the refractive-index change in the dark can be described by

$$n(t) = \left(\{ \Delta n_\mathrm{c} - \Delta n_\mathrm{b}^0 \cos\phi [\exp(-t/\tau_2)^\gamma] \}^2 + \{ \Delta n_\mathrm{b}^0 \sin\phi [\exp(-t/\tau_2)^\gamma] \}^2 \right)^{-\frac{1}{2}} , \tag{10}$$

where τ_2 is for the time constant for erasure of the grating Δn_b and γ is the exponent of a stretched exponential [30].

From the time evolution of Δn_b we can deduce that this grating is mainly caused by the photovoltaic effect. This is the dominating mechanism for charge redistribution in doped $LiNbO_3$. Because during thermal fixing the space-charge field is permanently compensated by screening ions that are

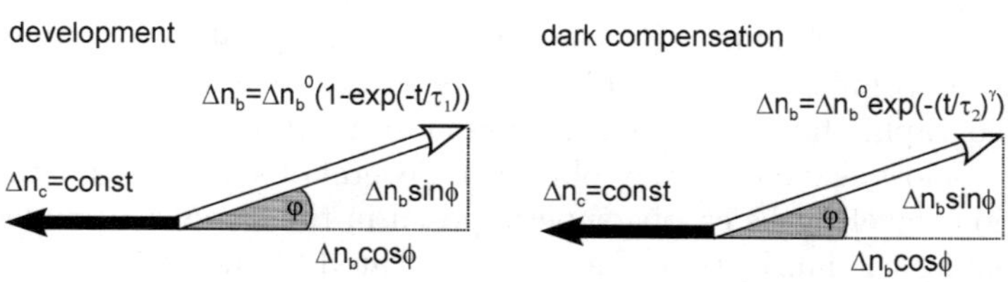

Fig. 10. Model for the refractive-index evolution during the development process and in the dark for reduced Cu-doped waveguides

mobile at high temperatures, a deep modulation of the Cu^+ traps takes place. Then, when the sample is cooled down to room temperature, a homogeneous light illumination generates modulated photocurrents, and the grating Δn_b appears. After the development process the photovoltaic grating Δn_b is slowly erased via the dark conductivity of the material, and the temporal development of this dark erasure can be described by a stretched exponential function [30]. Furthermore, a small diffusion grating is present that has a phase difference of 90° to the photovoltaic grating. This results in the small phase difference ϕ from the exact out-of-phase condition of 180° of Δn_b relative to the grating with amplitude Δn_c.

The permanent grating Δn_c is most likely created by a redistribution of Cu ions. This conclusion is based on the fact that the measured permanent gratings are uniquely found in samples that are doped with Cu and have never been observed in Fe-doped samples. This is also the reason why we can reject a modulated proton concentration as the origin of this permanent grating. We can also rule out that this effect results from an absorption grating formed during the recording process, because the efficiency of such a grating is much smaller (limited to $\eta < 7.2\%$ [25]) than that observed in our samples. There are two possible ways for the Cu ions to generate the grating Δn_c. First, the deep modulation of the Cu^+ and Cu^{2+} ions can already lead to material changes that cause a measurable refractive-index change and, second, it is not established that protons, as in $LiNbO_3$:Fe, are responsible for the compensation of the space-charge field during thermal fixing in $LiNbO_3$:Cu [21]. It may be possible that Cu ions move in the space-charge field during holographic recording and thus lead to material and density changes that may cause the observed additional refractive-index changes. This means that the Cu concentration itself will be modulated during fixing, not only the distribution of the Cu^+ and the Cu^{2+} ions alone.

5 Conclusions and Outlook

In this contribution we have reported on the formation of fixed reflection gratings in photorefractive Fe- and Cu-doped, Ti in-diffused $LiNbO_3$ channel waveguides for infrared light. Holographic gratings were recorded with visible light from an Ar ion laser at a high temperature of 180 °C. The gratings could be read with guided infrared light around 1.55 μm, and strong refractive-index modulations were obtained. By choosing the appropriate waveguide parameters and an optimization of the diffusion process, nearly polarization-independent holographic filters with efficiencies close to 100% and a bandwidth (FWHM) as low as 0.05 nm were obtained. Several of such filters with different reflection wavelengths can be superimposed in the same waveguide sample. This enables the fabrication of more complicated filters for DWDM, or for optical sensors that make use of the narrow spectral resonance. We have proven that it is possible to thermally adjust the peak wavelength of a reflection grating in a range of some tenths of nanometers, and the experimental verification of electrical switching of filters via the electro-optic effect is currently under investigation.

For commercial applications of thermally-fixed waveguide filters, or similar devices in photorefractive bulk samples that also require a fixation process, special care has to be taken to ensure the long-term stability of the holograms. In this sense, very promising results have been obtained for reduced Cu-doped $LiNbO_3$. In these samples, large refractive-index changes appear without the need of any additional development process. These strong refractive-index modulations do, most likely, not originate from the photorefractive effect but from material changes, i.e. density changes that are caused by the migration of Cu ions because of the electronic space-charge field. Such gratings are stable in the dark for at least one year with no degradation, and no compensation mechanism via dark conductivity is observed. On the other hand, a drawback of these samples is the increased absorption at 1.55 μm because of the strong Cu doping. However, we believe that these new results will stimulate the further development and application of devices based on photorefractive crystals in optical communications and integrated optics.

References

1. T. Tamir (Ed.): *Guided-wave Optoelectronics* (Springer, Berlin Heidelberg 1990)
2. A. M. Prokhorov, Y. S. Kuzminov: in E. R. Pike, R. G. W. Brown: *Physics and Chemistry of Crystalline Lithium Niobate* Hilger Ser. Opt. Optoelectron. (Adam Hilger, New York 1990)
3. A. M. Prokhorov, Y. S. Kuzminov, O. A. Khachaturyan: *Ferroelectric Thin-film Waveguides in Integrated Optics* (Cambridge Int. Science, Cambridge 1996)
4. R. V. Schmidt, I. P. Kaminow: Metal-diffused optical waveguides in $LiNbO_3$, Appl. Phys. Lett. **25**, 458–460 (1974)
5. J. L. Jackel, C. E. Rice, J. J. Veleka: Proton exchange in $LiNbO_3$, Ferroelectrics **50**, 165–170 (1983)
6. V. E. Wood, P. J. Cressman, R. L. Holman, C. M. Verber: Photorefractive effects in waveguides, in P. Günter, J.-P. Huignard (Eds.): *Photorefractive Materials and Their Applications II*, Topics Appl. Phys. **62** (Springer, Berlin, Heidelberg 1988)
7. D. Kip: Photorefractive waveguides in oxide crystals: fabrication, properties, and applications, Appl. Phys. B **67**, 131–150 (1998)
8. R. C. Alferness, R. V. Schmidt, E. H. Turner: Characteristics of Ti-diffused lithium niobate optical directional couplers, Appl. Opt. **18**, 4012–4016 (1979)
9. W. K. Burns, A. B. Lee, A. F. Milton: Active branching waveguide modulator, Appl. Phys. Lett. **29**, 790–792 (1976)
10. C. Becker, A. Greiner, T. Oesselke, A. Pape, W. Sohler, H. Suche: Integrated optical Ti:Er:$LiNbO_3$ distributed Bragg reflector laser with a fixed photorefractive grating, Opt. Lett. **15**, 1194–1196 (1998)
11. W. Sohler, H. Suche, Erbium doped lithium niobate waveguide devices, in E. Murphy (Ed.): *Design and Application of Integrated Optical Circuits and Components* (Marcel Decker, New York 1999)

12. F. S. Chen, J. T. LaMacchia, D. B. Fraser: Holographic storage in lithium niobate, Appl. Phys. Lett. **13**, 223–225 (1968)
13. A. Askin, G. D. Boyd, J. M. Dziedzic, R. G. Smith, A. A. Ballman, J. J. Levinstein, K. Nassau: Optically-induced refractive index inhomogeneities in $LiNbO_3$ and $LiTaO_3$, Appl. Phys. Lett. **9**, 72–74 (1966)
14. P. Günter, J.-P. Huignard (Eds.): *Photorefractive Materials and Their Applications I + II*, Topics Appl. Phys. **61**, **62** (Springer, Berlin, Heidelberg 1988)
15. D. Kip, B. Gather, H. Bendig, E. Krätzig: Concentration and refractive index profiles of titanium- and iron-diffused planar $LiNbO_3$ waveguides, Phys. Status Solidi A **139**, 241–248 (1993)
16. J. Hukriede, D. Kip, E. Krätzig: Copper diffusion into lithium niobate, Phys. Status Solidi A **172**, R3 (1999)
17. L. Hesselink, S. Orlov, A. Liu, A. Akella, D. Lande, R. Neurganonkar: Photorefractive materials for nonvolatile volume holographic data storage, Science **282**, 1089–1094 (1998)
18. J. Imbrock, D. Kip, E. Krätzig: Nonvolatile holographic storage in iron-doped lithium tantalate with continuous-wave laser light, Opt. Lett. **24**, 1302–1304 (1999)
19. J.J. Amodei, D. L. Staebler: Holographic pattern fixing in electro-optic crystals, Appl. Phys. Lett. **18**, 540–542 (1971)
20. H. Vormann, G. Weber, S. Kapphan, E. Krätzig: Hydrogen as origin of thermal fixing in $LiNbO_3$, State Commun. **40**, 543–545 (1981)
21. K. Buse, S. Breer, K. Peithmann, S. Kapphan, M. Gao, E. Krätzig: Origin of thermal fixing in photorefractive lithium niobate crystals, Phys. Rev. B **56**, 1225–1235 (1997)
22. J. Hukriede, D. Kip, E. Krätzig: Thermally fixed reflection gratings for infrared light in $LiNbO_3$:Ti:Fe channel waveguides, Opt. Lett. **23**, 1405–1407 (1998)
23. J. Hukriede, D. Kip, E. Krätzig: Thermal tuning of a fixed Bragg grating for IR light fabricated in a $LiNbO_3$:Ti channel waveguide, Appl. Phys. B **70**, 73–75 (2000)
24. J. Hukriede, D. Kip, E. Krätzig: Investigation of titanium- and copper-indiffused channel waveguides in lithium niobate and their application as holographic filters for infrared light, J. Opt. A **2**, 481–487 (2000)
25. H. Kogelnik: Coupled wave theory for thick hologram gratings, Bell Syst. Tech. J. **48**, 2909–2947 (1969)
26. P. M. Garcia, K. Buse, D. Kip, J. Frejlich: Self-stabilized holographic recording in $LiNbO_3$:Fe crystals, Opt. Commun. **117**, 235–240 (1995)
27. J. Frejlich, P. M. Garcia, A. A. Frechi: Deeply modulated stabilized photorefractive recording in $LiNbO_3$:Fe, Opt. Mater. **4**, 410–413 (1995)
28. K. Buse: Light-induced charge transport processes in photorefractive crystals: I models and experimental methods, Appl. Phys. B **64**, 273–291 (1997)
29. K. Peithmann, J. Hukriede, K. Buse, E. Krätzig: Photorefractive properties of lithium niobate volume crystals doped by copper indiffusion, Phys. Rev. B **61**, 4615–4620 (2000)
30. I. Nee, M. Müller, K. Buse, E. Krätzig: Role of iron in lithium niobate crystals for the dark-storage-time of holograms, J. Appl. Phys. **88**, 4282–4286 (2000)
31. J. Hukriede, D. Kip, E. Krätzig: Permanent narrow-band reflection holograms for infrared light recorded in $LiNbO_3$:Ti:Cu channel waveguides, Appl. Phys. B **72**, 749–753 (2001)

Optical Lambda-Switching at Telecom Wavelengths Based on Electroholography

Aharon J. Agranat

Department of Applied Physics, The Hebrew University of Jerusalem
Jerusalem 91904, Israel
Agranat@shum.huji.ac.il

Abstract. Electroholography is a wavelength-selective optical switching method based on governing of the reconstruction process of volume holograms by means of an electric field. Electroholography is based on the voltage-controlled photorefractive effect in the paraelectric phase. The basic switching device is an electrically controlled Bragg grating or a volume hologram stored in a volume of a paraelectric crystal by the photorefractive process.

The basic electroholographic switching operation is the reconstruction of a volume grating (hologram), which requires that the Bragg condition be satisfied, and therefore is wavelength selective. In addition the applied field governs the efficiency of the reconstruction. Consequently, electroholographic switching includes grouping, multicasting, power management and non-intrusive data as an integral part of the switching operation. In preliminary measurements the performance envelope of the electroholography-based switch, is a cube of 1.8 mm^3 was found to be as follows:

The minimum net insertion loss is 0.5 dB per switching operation. The minimum loss when a beam propagates through a latent grating is 0.2%. The Polarization-Dependent Loss (PDL) in a device that includes diversity architecture is less than 0.4 dB and the Polarization Mode Dispersion (PMD) is less than 0.07 ps. Bit-error rate (BER) in a switch operating at 40 Gb/s was measured to be 10^{-13}. These features make electroholography ideal for circuit switching applications. Finally, response times of approximately 10 ns were measured, opening the way to burst switching applications.

1 Introduction

Recent years have witnessed an exponential growth of the volume of Internet traffic, accompanied by significant improvements in the performance level and cost effectiveness of the communications technologies. In particular, in optical fiber communication systems, which are the dominant component of the communication infrastructure, these improvements were brought about by two major innovations: the erbium doped fiber amplifier, and the Wavelength Division Multiplexing (WDM) technology. Both innovations affected primarily the 'long haul' section of the network, enabling the data traffic capacity of the backbone of the optical networks to triple every eighteen

P. Boffi, P. Piccinin, M. C. Ubaldi (Eds.): Infrared Holography for Optical Communications,
Topics Appl. Phys. **86**, 129–156 (2003)

months. However, the growth rate of the performance of semiconductor devices that are currently used for communication routing has not evolved at a similar rate. Thus, barring a miraculous leap in the performance of silicon, it is clear that the Internet will soon need routing devices that simply cannot be built with the available electronic technology. In order to avoid this projected bottleneck, it is essential to develop an alternative routing technology. In particular, photonic switching where the routing is done in the optical domain is the obvious choice. Substantial effort is currently being devoted to develop a viable photonic switching scheme based on different technologies including microelectronic machining, liquid crystals, integrated optics, and piezoelectric mirrors.

In this paper we describe an alternative generic photonic switching scheme: electroholography (EH). EH is a generic beam-steering method based on the voltage-controlled photorefractive effect in the paraelectric phase. EH enables the governing of the reconstruction process of volume holograms by means of an externally applied electric field. As will be shown, the application of an electric field on paraelectric crystals in which volume holograms are stored as a spatial distribution of a space charge is a necessary condition for the reconstruction of these holograms. Thus, the application of an electric field can be used to activate prestored holograms that determine the routing of data-carrying light beams. The idea of EH was first proposed by *Agranat* as a generic concept for implementing optoelectronic artificial neural networks [1], and it was later extended as a generic method for spatial light modulation [2]. Indeed, the first demonstration of the EH concept was in the form of the EH neuron by *Balberg* et al. [3]. It was followed by a demonstration of a multistate EH switch and its use for the implementation of parallel multistage interconnection networks by *Pessach* et al. [4]. It is henceforth argued that the concept of EH forms the basis for a *wavelength-selective* photonic switching scheme, and as such is particularly suitable for routing in WDM networks.

2 The Electrically Controlled Bragg Grating

The basic building block of electroholographic wavelength-selective switching is the electrically controlled Bragg grating (ECBG) as shown schematically in Fig. 1.

When the electric field is off, the grating is in its latent state. In this state (Fig. 1a), the grating is transparent so that the incident lightwave propagates through the grating unaffected. When the electric field is turned on (Fig. 1b, the grating is activated. In the 'on' (active) state an input beam will be diffracted provided it fulfills the Bragg condition (the beam at wavelength λ_1 in Fig. 1b). An input beam that does not fulfill the Bragg condition will propagate through the active grating unaffected (the beam at wavelength λ_2 in Fig. 1b). Thus, electrically controlled gratings possess the basic features for

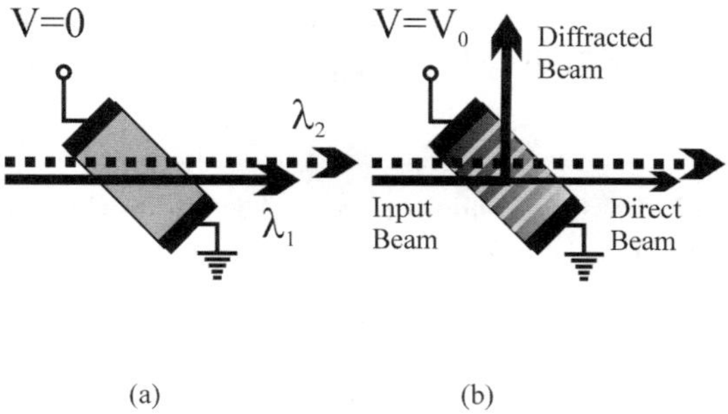

Fig. 1. The electrically controlled Bragg grating (**a**) in the latent state and (**b**) in the active state

functioning as wavelength-selective switches. In order to assess the functions that can be implemented by an electrically controlled grating, consider an ECBG in the transmission configuration as presented schematically in Fig. 2.

Assume first that the Bragg grating in Fig. 2 is passive, namely it is a grating of fixed refractive index modulation and period. The behavior of light beams propagating through a medium with periodic modulation of the (complex) index of refraction is derived from the coupled wave theory of *Kogelnik* [5]. (A concise and methodological presentation of the coupled-wave theory was published by *Solymar* and *Cooke* in [6]). Assume that the refractive index in the medium with the imprinted grating is given by

$$n = n_0 + \delta n \cos(Kx) , \tag{1}$$

where n_0 is the refractive index of the medium in which the grating was imprinted, δn is the amplitude of the Bragg grating and K is the grating vector, assumed co-aligned with the x axis so that the equiphase planes of the grating are perpendicular to the medium surface. A light beam incident

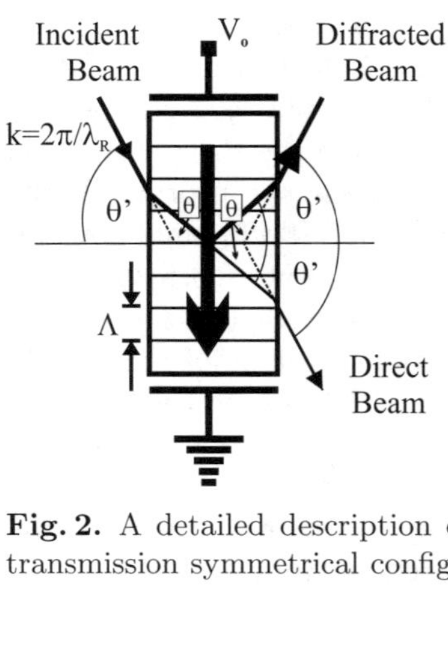

Fig. 2. A detailed description of the electrically controlled Bragg grating in the transmission symmetrical configuration

on the grating surface at an angle θ_B (given inside the medium) will be diffracted at an angle $-\theta_B$ provided it fulfills the Bragg condition given by

$$2n_0\Lambda\sin(\theta_B) = \lambda\,, \tag{2}$$

where λ is the wavelength of the light in the vacuum, and Λ is the period of the grating (namely: $\Lambda = 2\pi/|\boldsymbol{K}|$). The efficiency of the diffraction when the Bragg condition is satisfied is given by

$$\eta_{\text{diff}} = \sin^2(\kappa)\,, \tag{3}$$

where η is defined as the ratio between the intensities of the diffracted beam and the input beam, respectively, and κ is the coupling constant given by

$$\kappa = \frac{\pi\delta n d}{\lambda\cos(\theta_B)}\,, \tag{4}$$

where d is thickness of the grating. If the input beam does not fulfill the Bragg condition (2), the diffraction efficiency is given by

$$\eta_{\text{diff}} = \frac{\sin^2(\xi^2+\kappa^2)^{\frac{1}{2}}}{(1+\frac{\xi^2}{\kappa^2})^2}\,, \tag{5}$$

where ξ is the Bragg detuning factor given by

$$\xi = \delta\theta\beta d\sin(\theta_B)\,, \tag{6}$$

where $\delta\theta$ is the angular deviation from the Bragg angle θ_B and $\beta = 2\pi n_0/\lambda$ is the wave propagation factor inside the prism.

Assume now that the refractive index in the medium depends on the external field E applied to the medium, and is given by

$$n(x) = n_0(E) + rE\cos(Kx)\,. \tag{7}$$

The application of the field transforms the grating from a latent state in which $\delta n = 0$ to an active state in which $\delta n \neq 0$. In the former state $\eta = 0$, and hence an incident beam continues in its original direction unaffected regardless of its wavelength. In the latter state an incident beam that fulfills the Bragg condition (2) will be diffracted with efficiency η given by (3), whereas an input beam that deviates substantially from the Bragg condition will not be diffracted. Thus, a single ideal ECBG possesses in principle the basic feature of a wavelength-selective switching device. Two points however must be addressed up front: (i) The application of the external field affects the constant component of the index of refraction n_0 as well. This fact should be taken into account when designing the operation envelope of the switch, as it affects the Bragg condition. (ii) Normally, the ECBG switches less than 100% of the input power. This is due to imperfections in the grating, and the fact that the $\eta = 100\%$ point is a singular point of the applied voltage where $\kappa = \pi/2$. Therefore, a residual fraction of the input power continues in the original direction even when the grating is activated and may cause cross-talk

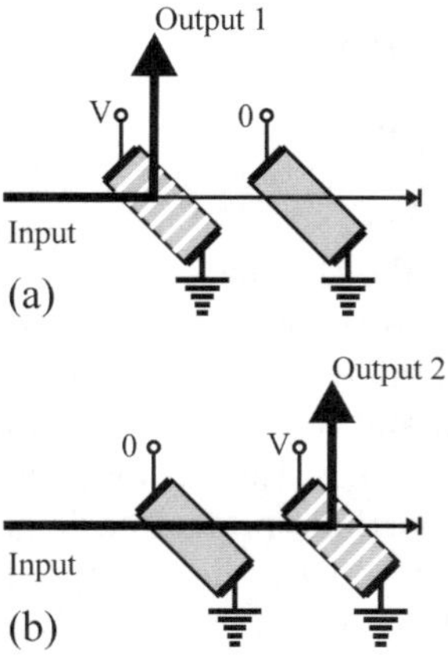

Fig. 3. A schematic description of a 1 × 2 electroholographic switch. (**a**) State 1: the input is connected to output 1, (**b**) State 2: the input is connected to output 2

between different data channels. The latter limitation is completely obviated when more than one ECBG is used in the switching device. Consider Fig. 3 presenting a schematic description of a 1 × 2 switch.

The switch is constructed of two ECBGs that are placed in series along the optical axis of the input beam. When the external field is applied to the first grating (Fig. 3a), this grating is activated causing the diffraction of the input beam to emerge out of output 1. Similarly, when the external field is applied to the second grating (Fig. 3b), the beam is diffracted off the second grating and emerges out of output 2. Several generic properties of the ECBG based switch should be noted.

- In both states of the switch the output beam is a diffracted beam, and hence the residual power that remains in the direct beam does not cause cross-talk. This residual power can be used to monitor the switch without interfering with its operation.
- The applied electric field governs the power of the diffracted beam. Therefore, the basic switching operation that is implemented by the ECBG is an analog operation. This fact enables the integration of power management function in the switch fabric itself.
- By governing the applied electric field to both ECBGs simultaneously, the power of the direct beam can be distributed between the two outputs. This property will be exploited later to provide multicasting capability.

In the next two sections the physical mechanism used for implementing the ECBG in photorefractive materials in the paraelectric phase will be presented.

3 The Physical Basis of Electroholography

3.1 The Voltage-Controlled Photorefractive Effect in the Paraelectric Phase

The term PR (photorefractive) effect is used to describe the phenomena related to the light-induced creation of spatial changes in the refractive index. It is normally attributed to the formation of a metastable space charge that is spatially correlated with the incident illumination, and induces changes in the index of refraction through the electro-optic effect. Thus, the PR medium is electrooptic and contains a layer of traps that are partially populated (e.g. iron-doped $LiNbO_3$ in which the Nb^{+5} ion is substituted by the Fe^{+3} and Fe^{+2} ions, the latter serving as the empty and populated traps respectively). A qualitative description of the photorefractive process, which leads to the formation of a sinusoidal index grating, is presented in Fig. 4.

Two plane waves interfere in the photorefractive medium creating a sinusoidal interference pattern (Fig. 4a). This causes photoionization of the populated traps, and consequently a generation of free charge carriers at a rate that is spatially correlated with the interference pattern (Fig. 4b). The free charge carriers are then transported by diffusion, drift or the bulk photovoltaic effect, and are eventually retrapped by the empty traps. The outcome of this process is a trapped space charge that is spatially correlated with the exciting illumination interference pattern (Fig. 4c). The electric field induced by the space charge induces a modulation in the index of refraction through the electro-optic effect. In crystals in the ferroelectric phase the electro-optic

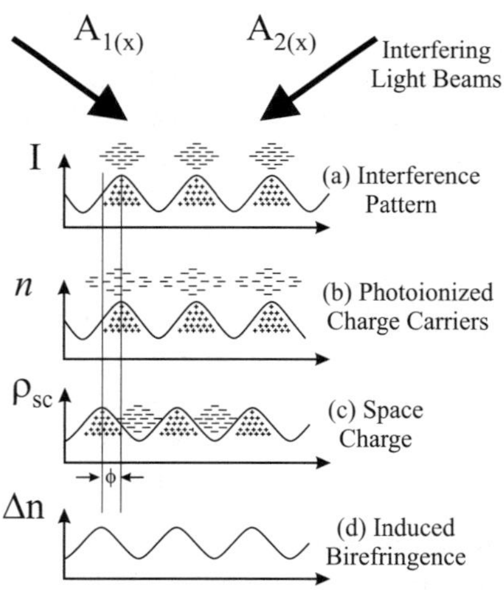

Fig. 4. A qualitative description of the photorefractive process. (**a**) Formation of the interference pattern. (**b**) Generation of the free charge carriers. (**c**) The distribution of the trapped space charge. (**d**) The spatial modulation of the index of refraction (birefringence)

effect is *linear*. Therefore, the modulation of the index is spatially correlated with the space-charge field, and hence with the interference pattern of the excited illumination (Fig. 4d). (The PR effect has been discussed at length in several monographs and papers, cf. [7,8,9,10]).

In crystals in the paraelectric (PE) phase the electro-optic effect is quadratic and hence the induced change in the index of refraction is given by

$$\Delta n = \tfrac{1}{2} n_0^3(E) g_{\text{eff}} P^2 , \tag{8}$$

where Δn is the induced birefringence, n_0 is the refractive index, g_{eff} is the effective quadratic electro-optic coefficient and P is the dc (or low frequency) induced polarization [11]. (Note that (8), which is generally a tensor equation, is written here in its scalar form with the appropriate *effective* quadratic electro-optic coefficient [11].) Consider first the case of a transmission sinusoidal grating (Fig. 5).

The grating is formed by the interference of two plane waves with wave vectors $\boldsymbol{k}_1$ and $\boldsymbol{k}_2$ respectively ($|\boldsymbol{k}_1| = |\boldsymbol{k}_2| = k$), incident symmetrically with respect to the normal to the crystal surface. The light intensity of the interfering beams during the writing of the hologram is given by $I = I_o + I_1 \cos(Kr)$, where $\boldsymbol{K}$ is given by $\boldsymbol{K} = \boldsymbol{k}_1 - \boldsymbol{k}_2$. During the PR process this interference pattern generates a space charge given approximately by $E_{\text{sc}}(r) = E_{\text{sc}} \cos(\boldsymbol{K} r + \phi)$. If an external electric field E_{R} is applied to the crystal, then the electric field in the crystal is given by

$$E = E_{\text{R}} + E_{\text{sc}}(\boldsymbol{r}) . \tag{9}$$

In the PE phase the induced polarization P is in the linear region, namely:

$$P = \epsilon_0(\epsilon_{\text{r}} - 1)E \approx \epsilon_0 \epsilon_{\text{r}} E , \tag{10}$$

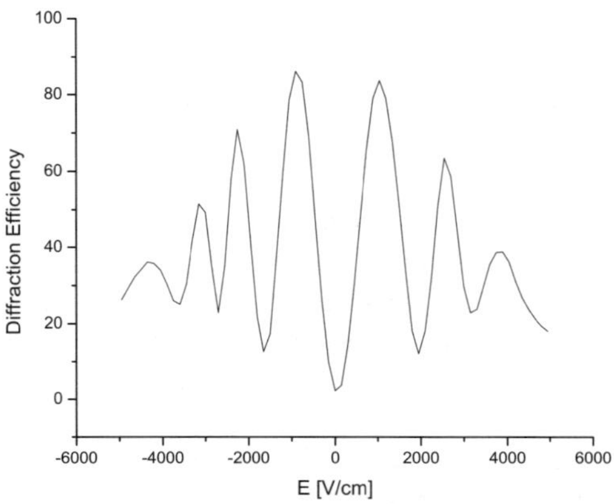

Fig. 5. Experimental results of the diffraction efficiency vs. the electric field of a Bragg grating in the transmission symmetrical configuration in KLTN

where ϵ_0 is the dielectric permittivity, ϵ_r is the relative dc dielectric constant, and it is assumed that the crystal is slightly above T_c so that $\epsilon_r \gg 1$. Since the electro-optic effect is quadratic, the induced birefringence is given by

$$\Delta n = \frac{1}{2} n_0^3 g_{\text{eff}} \epsilon_0^2 \epsilon_r^2 [E_R + E_{sc}(\boldsymbol{r})^2] \ldots$$
$$= \frac{1}{2} n_0^3 g_{\text{eff}} \epsilon_0^2 \epsilon_r^2 [E_R^2 + E_{SC} r^2 + 2E_R E_{SC}(r)] \,. \tag{11}$$

Thus the induced birefringence will contain three terms: (i) a constant shift proportional to $(E_R)^2$, (ii) a linear grating proportional to $E_{sc}(r)$ and (iii) a quadratic grating proportional to $[E_{sc}(r)]^2$. The effect of the induced birefringence given by (11) on an incoming light beam with wavelength λ_R, entering the crystal with an internal angle of incidence q_0, which fulfills the Bragg condition (2) for a grating with period $\Lambda = 2\pi/K$, is as follows: the term proportional to $(E_R)^2$ does not contain spatial information but introduces a constant shift to the index of refraction. This term may cause a violation of the Bragg condition. However, if the incoming beam enters the crystal at an angle θ' (Fig. 2), then Snell's law requires that $\sin\theta' = n_0 \sin\theta_B$. Hence, the Bragg condition is not violated in the transmission configuration and this term does not affect the diffraction. The term proportional to $[E_{sc}(r)]^2$ is a quadratic grating for which the grating vector is 2 K. Hence, the quadratic grating does not contribute to the diffraction of the incoming beam since the Bragg condition is not fulfilled. The term proportional to $E_0 E_{sc}(r)$ induces a grating for which the incoming beam fulfills the Bragg condition. The amplitude of this grating $\delta(\Delta n)$ is given by

$$\delta(\Delta n) = n_0^3 g_{\text{eff}} \epsilon_0^2 \epsilon_r^2 E_R E_{sc}(r) \,. \tag{12}$$

This grating will cause diffraction of the incoming beam with a diffraction efficiency given by (3). Thus, the power of the switched beam will be given by

$$P_{\text{diff}} = P_{\text{in}} \exp(-\alpha d) \sin^2\left(\frac{\pi n_0^3 g_{\text{eff}} \epsilon_0^2 \epsilon_r^2 E_R E_{sc}(r) d}{\lambda_R \cos(\theta)}\right), \tag{13}$$

where P_{in} is the power of the incoming beam, α is the absorption coefficient of the crystal, d is the crystal thickness and $\delta(\Delta n)$ is given by (12). In summary, the information-carrying space-charge field is transformed into a modulation of the refractive index only in the presence of an external electric field. Therefore, the use of the quadratic electro-optic effect enables an analog control of the efficiency of the reconstruction of the information. This phenomenon is known as the voltage-controlled PR effect [12]. It should be noted that the effect of the electric fields on PR gratings was investigated in ferroelectrics such as SBN, where the effect of the applied field on the diffraction is achieved by the Bragg detuning mechanism [13].

3.2 The Voltage-Controlled Photorefractive Effect in KLTN

The implementation of EH-based devices (EHD) necessitated the development of a special photorefractive crystal: potassium lithium tantalate niobate (KLTN) [14]. As pointed out above, the optimal work point for the electroholographic device is above the Curie point, where the Curie–Weiss law holds and the induced polarization is given by (10). At the same time it is desirable to approach the Curie point as much as possible so that $\epsilon_{\mathrm{r}} \approx 10^4$ and large polarizations can be induced with moderate fields.

The first demonstration of the voltage-controlled PR effect was done in potassium tantalate niobate (KTN) doped with copper and vanadium [12]. KTN is a ferroelectric oxide in which large photorefractive effects were demonstrated. The Curie temperature T_{c} of KTN is determined by the ratio Nb/Ta at the rate of $\Delta T_{\mathrm{c}} \approx 8.5\%$ per mole of Nb in the crystal [15]. Thus, controlling this ratio during the growth of the crystal enables setting the work point to the desirable point. In KTN crystals in which the concentration of Nb exceeds 30% per mole, the optical quality is substantially degraded in the proximity of the phase transition. Therefore, KTN is not the optimal medium for devices in which the desirable work point is approximately at room temperature. This limitation was obviated in KLTN which is a derivative of KTN [16,17]. It was found that the addition of Li to $\mathrm{KT}a_{1-x}\mathrm{Nb}_x\mathrm{O}_3$ (KTN) causes an increase in the ferroelectric phase transition temperature T_{c}, while maintaining high optical quality in the temperature region close to the transition. It was found that KLTN doped with copper and vanadium is particularly suitable for the implementation of EHDs. The PR process in this material is produced by photoexcitation of electrons from Cu^{+} ions by photons with energy exceeding 2 eV that are transported and eventually trapped by Cu^{++} ions.

High-quality KLTN crystals were grown using the top seeded solution growth (TSSG) method [18]. By adjusting the Nb/Ta and Li/K ratios in the flux it was possible to set T_{c} in the range 100 K to 400 K while maintaining high optical quality in the region close to the transition. In particular, KLTN crystals doped with copper and vanadium were grown with $T_{\mathrm{c}} \approx 290\,\mathrm{K}$, in which the optical quality in the region slightly above T_{c} (i.e. in the region $T \geq T_{\mathrm{c}} + 4\mathrm{K}$) was especially good.

A typical measurement of the diffraction efficiency of a symmetrical transmission grating (Fig. 2) as function of the electric field is given in Fig. 5 [19]. The crystal was potassium lithium tantalate niobate (KLTN) doped with copper and vanadium with $T_{\mathrm{c}} = 18.5\,°\mathrm{C}$. A sample of dimensions $1.5\times1.5\times2\,\mathrm{mm}^3$ was cut along the crystallographic axes. Gold electrodes were deposited orthogonal to the optical axis. The grating was written at 532 nm with an intensity of approximately 10 mW per writing beam. The exposure time was 15 s. The grating was written and read at $T = 25\,°\mathrm{C}$.

Note that in a symmetrical transmission grating the Bragg condition is compensated by the Snell refraction at the entrance plane. Therefore, the diffraction efficiency is expected to follow (13). However, as can be seen in Fig. 5 the dependence of η_{diff}. on the applied field deviates from (13) as the strength of the applied field increases. This is due primarily to the saturation of the induced polarization that occurs in the vicinity of the phase transition [20].

3.3 Assessment of the KLTN Crystal as an Electroholographic Medium

In order to assess the performance limits of the KLTN-based electroholographic devices it is necessary to analyze the behavior of the KLTN crystal as a holographic storage medium. The latter is manifested by several parameters that will henceforth be considered.

3.3.1 The Holographic Storage Capacity

The material property that limits the storage capacity in a holographic system is the maximum induced change in the refractive index that can be generated in the holographic storage medium. One of the critical parameters that govern the performance envelope of an EHD is the maximum diffraction efficiency of the grating. In a die of given dimensions, the latter is also limited by the maximum induced change in the refractive index that can be generated by the EH process. Therefore, the holographic storage capacity is a useful parameter for comparing the potential performance of EH media. *Mok* et al. [21] suggested that the storage capacity of a holographic system could be characterized by the figure of merit $M/\sharp$ defined as

$$M/\sharp = M\sqrt{<\eta>}\,, \tag{14}$$

where $<\eta>$ is the average diffraction efficiency of the stored holograms. It is shown that for $M \gg 1$ the $M/\sharp$ is independent of M; it does, however, depend on the angular and wavelength configuration of the reading system. Detailed measurements of the $M/\sharp$ of KLTN are described and discussed in [22]. The $M/\sharp$ of copper- and vanadium-doped KLTN in the configuration of an EHD, where the reading wavelength is in the range of the WDM wavelengths (approximately 1550 nm $\pm$ 100 nm), and the (external) angle between the input beam and the switched beam is 90°, were performed recently by *Puterkovsky* [23]. $M/\sharp = 2.13$ was obtained in a 1.65 mm thick sample, operating at $T_c + 5\,°\mathrm{C}$ under applied field of 3.6 kV/cm. This result is equivalent to $M/\sharp \approx 20 \in$ the same sample operating under the same conditions with a readout wavelength of 500 nm. Note that this result is exceptionally high compared to previous results obtained in both $LiNbO_3$ and KLTN as reported in references [21] and [22], respectively.

3.3.2 The Photorefractive Sensitivity

The light energy required to generate the space charge grating during the fabrication process of the EHD is governed by the PR sensitivity S. S is defined as the generated change in the index of refraction per absorbed light energy in a unit volume of the material. In terms of S, the dependence of the change in the refractive index generated by the PR process on the exposure time τ and on the intensity of the writing beams I_0 is given by

$$S = \frac{\delta(\Delta n)}{I_W \alpha_W \tau}, \tag{15}$$

where α_W is the absorption of the EH medium at the writing wavelength. The PR sensitivity in copper- and vanadium-doped KLTN was measured recently by *Puterkovsky* [23]. It was found that at wavelengths around $\lambda \approx 500\,\text{nm}$, $S \approx 1.3 \times 10^{-4}\,\text{cm}^3/\text{J}$. Thus, in order to generate a grating with amplitude of $\delta(\Delta n) = 10^{-4} \in$ an EH medium with $\alpha_W = 1\,\text{cm}^{-1}$, it is required to use $I_W = 1\,\text{W}$ for approximately $1\,\text{s}$.

3.3.3 Holograms Stability

The sustained stability of electroholograms over long periods of time under operating conditions is essential for the establishment of electroholography as a viable technology. The stability of electroholograms implemented by the PR process is determined by the stability of the space charge that forms the electroholograms in their latent state. In principle, if free charge carriers are generated in the EH medium, they will tend to restore thermodynamic equilibrium by migrating under the influence of the space-charge field in a direction that causes the erasure of the space charge. Two mechanisms may cause the generation of charge carries in the electroholographic medium: optical excitation and thermal excitation. Both mechanisms may lead to the decay of the latent electroholograms, and therefore it is imperative to quantify their effect.

Optically Induced Decay of the Holograms. Electroholograms produced by the PR process are by definition subject to erasure during readout [24]. Recall that in its latent form the EH grating is a space-charge grating formed by the excitation of trapped charge carriers that are transported and retrapped. Illuminating a PR grating will excite charge carriers that will drift by the influence of the space-charge field, and will cause the decay of the latter. Thus, illumination at a wavelength at which the PR sensitivity is zero will not cause erasure during readout since the illuminating photons do not possess enough energy to excite the charge carriers. It is therefore required that at the operating wavelengths the PR sensitivity of the PR medium used in the device will be zero. It is further required that the PR medium in the

EHD will not be exposed to stray light after the writing of the grating except for the lightwave emanating from the fiber. In KLTN crystals doped with copper the PR process is produced by the redistribution of electrons between Cu^+ and Cu^{++} ions. The energy required to photoionize Cu^+ ions exceeds 2 eV. Hence, it is not expected that the photons at the operating wavelength ($h\nu \geq 1.25\,\mu m$) will have sufficient energy to cause photoexcitation of electrons from the Cu^+ ions. Measurements of the absorption and photoconductivity in copper-doped KLTN indicate that the PR sensitivity in the wavelength range of the DWDM bands is indeed zero. Therefore, optical erasure of the gratings is not expected to happen in this wavelength range.

Thermal Decay of Electroholograms. Thermally excited charge carriers are expected to cause erasure over time of the space charge that forms the electrohologram. We distinguish between two possible mechanisms of thermal excitation: the intrinsic mechanism and the extrinsic mechanism. The intrinsic mechanism relates to the thermal excitations of the charge carriers that form the PR space charge. In copper-doped KLTN two intrinsic thermal decay processes are possible. In the first an electron is thermally excited from a Cu^+ state, drifts under the space-charge field, and is eventually captured by a Cu^{++} state. The second process is the thermal release of a hole from a Cu^{++} state to the valance band, and its subsequent capture in a Cu^+ state. The probability for the thermal emission of a charge carrier is given by $\nu_0 \exp(-\mathrm{E_a}/\mathrm{k_B}T)$, where ν_0 is the 'attempt to escape' frequency, and E_a is the energy barrier that needs to be surmounted. In KLTN, a Cu^+ level is located more than 2 eV below the conduction band edge, and the Cu^{++} states are located a similar energetic distance away from the valance band edge. Even for an overestimation of the 'attempt to escape frequency' (10^{20} Hz) the timescale for the thermal release of carriers is over 1000 years. Hence, erasure of the electrohologram due to intrinsic thermal decay of the space charge can be discarded. The second mechanism is the extrinsic mechanism. The PR medium may contain localized states other than those that carry the PR space charge. These states can reside close to the band gap edges of the material, so that their activation energy is substantially less than that of the PR states. Thermal excitation of charge carriers from such states will cause the decay of the space charge forming the latent electrohologram, and hence the erasure of the latter. The extrinsic mechanism can be obviated by removing the extrinsic states either during the crystal growth or by some treatment after the growth.

The lifetime of the electroholograms is defined as the time over which diffraction efficiency is reduced to 90% of its original value under the operating conditions. Very long lifetimes are measured through the acceleration of the thermal relaxation processes by controlled heating of the electroholograms. Using this method, it was found that in copper-doped KLTN, electroholograms that were subjected to the process that removes the extrinsic states have lifetimes of over ten years.

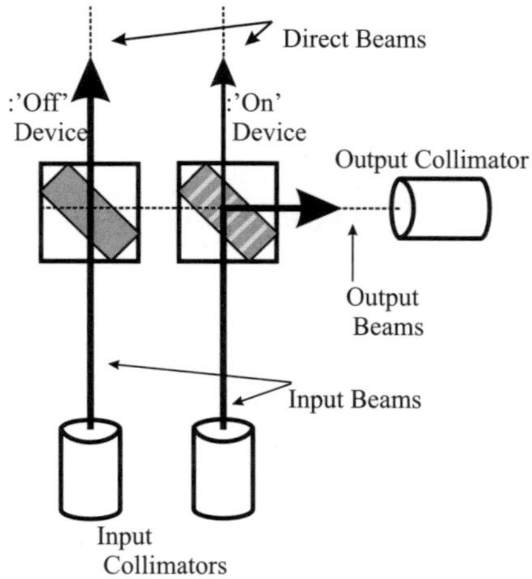

Fig. 6. The basic electroholographic device in the latent and active states

4 The Basic Electroholographic Switch Module

The basic EH-based switch *module* (EHSM) is an array of single EH-devices. The basic single EH-based device is an ECBG tuned to a single-wavelength on the International Telecommunication Union (ITU) grid as shown schematically in Fig. 6.

4.1 The Architecture and Operation of the Basic EH Switch Module

The input signal is incident on the grating after it is was collimated by the input collimator. The period of the grating is chosen so that the diffracted beam emerges from of the device at 90° to the input beam. Consider the operation of the device when the input signal contains one component (wavelength) of the Dense Wavelength Division Multi Plexing (DWDM) signal. When the ECBG is in the 'off' state, the input beam propagates unaffected through the device. When the ECBG is in the 'on' state (the electric field is applied), the input beam is diffracted and emerges from the device and is collected by the output collimator. As the efficiency of the diffraction is not 100%, a residue of the input beam emerges directly from the device and continues along the input optical axis.

In the EH-based switch module the EHDs are arranged in a 'full crossbar' configuration in which the columns are inputs and the rows are outputs. Each column in the cross-bar is allocated a wavelength on the ITU grid. Namely, the EHDs that constitute the column are tuned to the same wavelength. The output (diffracted) beams that propagate along the optical axis of a row are coupled by one output collimator into an output fiber. The input WDM signal entering the device through the input port is first demultiplexed into its single

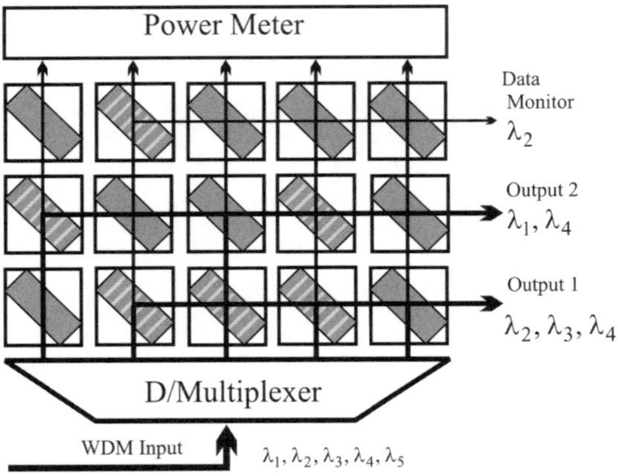

Fig. 7. An electroholographic switch module performing grouping, multicasting, power management and data monitoring

(one-wavelength) components. Each of these components is then collimated and directed along the optical axis of its respective column. By activating one of the EHDs the respective component will be diverted to the selected output. An EHSM constructed of five input columns and three output rows is shown schematically in Fig. 7.

The various functions implemented by the EHSM are demonstrated in Fig. 7 and are described as follows.

4.1.1 Grouping

The EHSM can regroup the input WDM channels into subgroups and assign each subgroup to an output port. (In Fig. 7 the group $\{\lambda_2, \lambda_3, \lambda_4\}$ is assigned to output 1, and group $\{\lambda_1, \lambda_4\}$ is assigned to output 2. Note that the 'grouping' function is in fact an extension of the basic 'dynamic drop' function that describes the switching of one WDM lightwave channel to an output port. It can also be used to perform the 'add' function as illustrated schematically in Fig. 8.

The 'add' WDM channels are directed along the input row and are selected by the EHD of the respective output row. (In Fig. 8, $\{\lambda_1, \lambda_2, \lambda_4, \lambda_5\}$ are carried by the input, $\{\lambda_2, \lambda_5\}$ are directed to output 1, and $\{\lambda_1, \lambda_4\}$ are directed to output 2.)

4.1.2 Multicasting

The EHSM can multicast the power of each of the input WDM lightwave channels between several outputs. Multicasting is accomplished by controlling the level of the applied electric field that governs the diffraction efficiency of the EHDs. (In Fig. 7 λ_4 is distributed between outputs 1 and 2.)

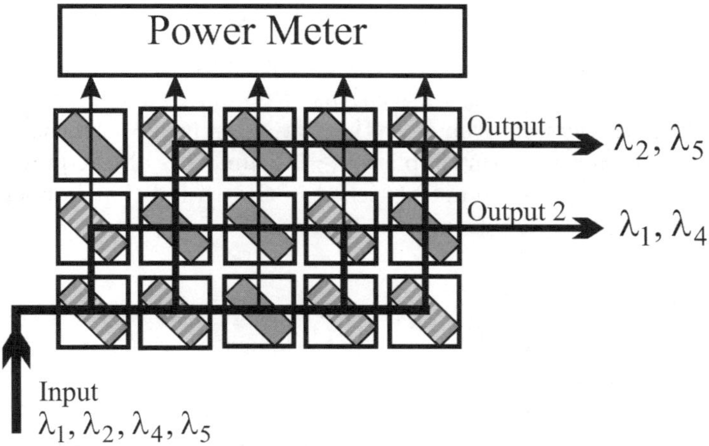

Fig. 8. An electroholographic switch module performing the 'add' operation

4.1.3 Power Management

The EHSM enables the management of the power of the WDM channels in each of the subgroups. Power management is accomplished in three steps: (i) Monitoring the residual power that continues through the column along the input optical axis; (ii) estimating the switched power by assuming that the efficiency of the switching operation is known; and (iii) governing the switching efficiency by controlling the level of the applied electric field.

4.1.4 Data Monitoring

The EHSM enables monitoring of the data of the WDM channels without interfering with the switching operation. In principle, data monitoring can be accomplished by sampling the residual power that continues through the column along the input optical axis. This, however, would require the placement of a dedicated receiver at the top of each column. If continuous monitoring of all the channels in parallel is not required, sparse sampling can be accomplished by the 'data monitoring' row that diverts the selected channel to one receiver. (In Fig. 7, $lambda_2$ is diverted to the data monitor.)

4.2 The Performance Parameters of the Basic EH Switch Module

The basic building blocks of EH-based systems are the generic EHDs. The latter are grouped together to form EHSMs of different architectures according to the specific application of the EH-based system. In order to assess the performance envelope of EH-based systems it is necessary to evaluate the primitive performance parameters of the generic EHDs.

4.2.1 Operating Wavelengths Range

The EHD can be operated at wavelengths that do not affect the PR grating. Namely, the photons of the operating wavelengths do not have enough energy to photoionize charge carriers that can erase the grating. In KLTN crystals in which the PR impurity is copper this implies $\lambda \geq 1.25\,\mu\text{m}$. It is also required that the operating wavelengths are not absorbed in the crystal that implies in copper doped KLTN: $\lambda \leq 1.7\,\mu\text{m}$.

4.2.2 Data Throughput Rate

In principle, the data carried by light beams that propagate through an EHD is not affected by the diffraction that occurs in an active switch, provided the light is linearly polarized along a principal axis of the switch. In reality the polarization of the lightwave that emerges from the fiber is not linearly polarized and varies over time. The bit-error rate at very high data throughput rates will therefore be affected by the polarization mode dispersion and polarization dispersion loss that occurs in the switch.

4.2.3 Insertion Loss

Consider the path along which the lightwave propagates through an EHSM (Figs. 6, 7 and 8). The light propagating in the input fiber enters the module from the input collimator as a collimated beam. It then propagates through a series of passive EHDs along the input column until it reaches the active switch where it is diffracted along the output row where it propagates through passive switches. Finally, it is collected by the output collimator into the output fiber. The insertion loss in the EH switch module is given by

$$IL = IL_{\text{collimators}} + N_{\text{ps}} IL_{\text{ps}} + IL_{\text{as}}\,, \tag{16}$$

where

- $IL_{\text{collimators}}$ is the insertion loss due to the coupling between the input and output collimators. $IL_{\text{collimators}}$ takes into account the divergence of the beam in the module and is therefore affected by the length of the path traversed by the beam. $IL_{\text{collimators}}$ also depends on the alignment procedure of the collimators. For example, a beam with a diameter of 0.6 mm traversing through a path length of 7 cm and collimated so that the waist of the beam is midway between the collimators will cause an insertion loss of 1.5 dB.
- IL_{ps} is the insertion losses in one EHD at the passive mode. The bulk KLTN crystal does not absorb light at the operating wavelengths. Therefore, IL_{ps} is determined primarily by the loss of reflection, scattering and absorption at the input and output surfaces of the switch. It is also affected by the accuracy of the die parallelism achieved in the processing. It is expected that high-quality polishing and antireflection coating procedures will result in $IL_{\text{ps}} = 0.3\%$.

- N_{ps} the number of passive switches the beam passes through. Note that Nps depends on the specific route of the beam in the module. Consider for example the 5×2 module illustrated in Fig. 7. The data channel at λ_1 that is switched to output 2 goes through the longest route ($N_{ps} = 5$), whereas the data channel at λ_5 that is switched to output 1 goes through the shortest route ($N_{ps} = 0$).
- IL_{as} is the insertion loss in the active switch given by

$$IL_{as} = IL_{ps} + 10 \log_{10}(1 - \eta_{diff}) , \tag{17}$$

where η_{diff} is the diffraction efficiency in the active switch.

4.2.4 Polarization Dependence

The state of polarization of the lightwave signal emerging from the fiber into the switch is different from the one entering the fiber at the transmitter side and varies with time according to various environmental conditions such as temperature and stress. Thus, the state of polarization of the lightwave entering the switch fluctuates with time constants ranging from subseconds to hours according to the extent and rate of change of the said environmental conditions [25]. In order to avoid loss of data it is required that the switching operation does not depend on the state of polarization of the incoming lightwave signal. The effects of variations in the state of polarization of the propagating signal are quantified by the polarization-dependent loss (PDL), and the polarization mode dispersion (PMD). The PDL quantifies the loss of power of the switched lightwave signal as a function of its state of polarization. The PMD quantifies the time delay between pulses that are launched into the switch with orthogonal states of polarizations. A switching operation that is implemented in a crystal by diffraction from a photorefractive grating is in principle polarization-dependent. This is due to the fact that the electro-optic tensor that governs the diffraction efficiency is by definition polarization dependent [11]. Minimization of the PDL and PMD can be achieved at the device level by the polarization-independent configuration of the EHD, and at the module architecture level by polarization diversity. Polarization independent configuration of the switch is accomplished by exploiting the symmetry relations between the elements of the electro-optic tensor of the EH medium. This is done by selecting the direction of propagation of the lightwave, the grating vector and the applied field with respect to the crystal principle axes so that the diffraction efficiency is independent of the state of polarization. As an example consider the EHD with reflective configuration illustrated in Fig. 9.

Here the optical axis of the switch coincides with a principal axis of the KLTN crystal and the grating vector of the EH grating. Applying the electrical field along the optical axis causes the optical axis to become the axis of symmetry of a uniaxial crystal. Thus, an input beam propagating along

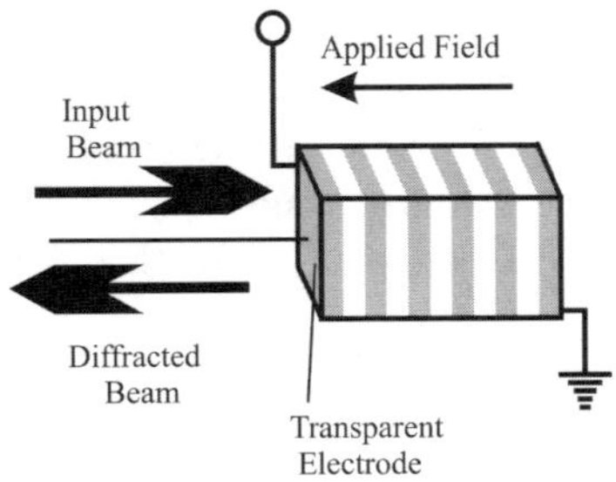

Fig. 9. An electroholography-based switch device in the reflection configuration which is PDL- and PMD-free

the optical axis will be diffracted in the opposite direction with a diffraction efficiency independent of its state of polarization. Note that in the reflective configuration the Bragg detuning is not compensated by the Snell refraction at the crystal input plane. Therefore, the diffracted efficiency is given by (5).

In polarization diversity schemes the two orthogonal states of polarization of the incoming lightwave signal are switched separately and then reunited. Thus the single EHD in a EHSM consists of two EH crystals, each switching one polarization branch of the same WDM component of the lightwave signal. Consider Fig. 10 which demonstrates an optional implementation of polarization diversity in one route of the EHSM.The polarizing beam splitter PBS1 at the input separates the incoming collimated lightwave signal into its two orthogonal states of polarization. Assume that maximum efficiency is achieved when the polarization of the lightwave signal is in the plane of the device. Branch 1 (the dashed line in Fig. 10) is switched by the switching device EHD1 and then its polarization is rotated by 90° by the $\lambda/2$ retardation plate RP1. The polarization of Branch 2 (the dotted line in Fig. 10) is first rotated by RP2 to the plane of the device and then switched by EHD2. The two branches are then united in the polarizing beam splitter PBS2. Thus both branches are switched with maximum efficiency. Note that the path length of both branches is the same. Preliminary measurements of the PDL vs. the wavelength were measured in a EHSM similar to the one illustrated in Fig. 10. The grating was written to operate at $\lambda_o = 1549.8\,\mathrm{nm}$. It was found that PDL $(\lambda_o) < 0.07\,\mathrm{dB}$ and PDL $(\lambda \pm 0.5\,\mathrm{nm}) < 0.4\,\mathrm{dB}$. In general it was found that

$$\frac{\partial PDL(\lambda)}{\partial \lambda} \approx 0.75\,\mathrm{dB/\,nm}\,. \tag{18}$$

The PMD was also measured in this module and was found to be approximately 0.07 ps at a data throughput rate of 10 Gb/ s. The low PMD in this module is also manifested in bit-error rate tests. A lightwave signal at a data throughput rate of 40 Gb/ s incident on the module with a power of 8 dB m yielded a BER $\leq 2 \times 10^{-13}$.

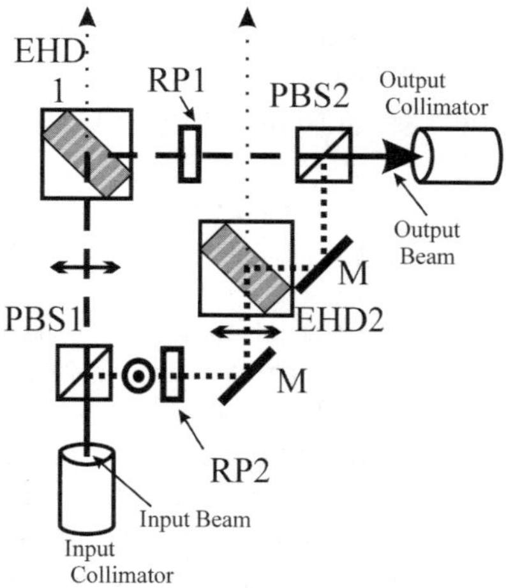

Fig. 10. An element of the electroholographic switch module in which an implementation of the polarization diversity architecture is shown

It should be noted that in an EHSM in which polarization diversity is applied it is possible to perform polarization management. The two polarization branches of each WDM component of the lightwave signal can be monitored separately by the direct beams propagating through the two crystals that in this case constitute the EHD. Combining this feature with the fact that the power of the switched (output) lightwave signal is governed by the applied electric field enables intelligent control of each of the polarization components of the signal. This feature may become useful in systems where correction of the polarization of the lightwave signal in real time is required.

4.2.5 Selectivity

The EHD is optimized to function as the basic building block in wavelength-selective switching systems. Therefore, the spectral response of the EHD is of paramount importance for the performance envelope of any EH-based system. The single switching EHD is in essence an electrically controlled filter. The spectral response of the EHD is derived from the effect of deviating from the Bragg condition (2). Consider an EHD implemented by an electrically controlled sinusoidal Bragg grating. The maximum switching efficiency of the EHD is obtained when the incident beam fulfills the Bragg condition (2). When the Bragg condition is not fulfilled the diffraction efficiency will diminish. The level of deviation from the Bragg condition is given by the Bragg detuning factor ξ. If the wavelength of the incident beam is shifted by $\delta\lambda$, ξ will be given by

$$\xi = \frac{\pi}{\Lambda} \frac{\delta\lambda d}{n\Lambda \cos(\theta_{\mathrm{B}})} , \tag{19}$$

assuming all other parameters remain fixed. In terms of ξ the diffraction efficiency diminishes according to expression (5). Deviation from the Bragg condition will also cause the diffracted (output) beam to divert from its original direction ($-\theta_{\mathrm{B}}$) by

$$\delta\theta_{\mathrm{out}} = \frac{\delta\lambda}{2n\Lambda\cos(\theta_{\mathrm{B}})} \,. \tag{20}$$

The spectral response of the EHD results from the combined effect of the loss of diffraction efficiency and the diversion of the output beam from its original direction. Consider the configuration of the EHD presented schematically in Fig. 6. The output collimator is aligned in a fixed position along the Bragg angle of the output beam. Diversion of the output beam from the Bragg angle as a result of the shift in the wavelength will reduce the power that is collected by the output collimator. The combined effect of the loss of diffraction efficiency and the diversion of the output beam from the Bragg angle on the spectral response of an EHD is demonstrated in Fig. 11.

A grating of thickness $d = 1.5\,\mathrm{mm}$, with period of $\Lambda = 1.1\,\mathrm{mm}$ was stored in a KLTN crystal ($n_0 = 2.3$). The Bragg condition is fulfilled for an input Gaussian beam at $\lambda = 1.55\,\mu\mathrm{m}$ with diameter of $0.510\,\mathrm{mm}$ incident at 45° to the plane of incidence of the crystal. The solid line in Fig. 11 presents the power diffracted from the crystal and detected by a wide area detector

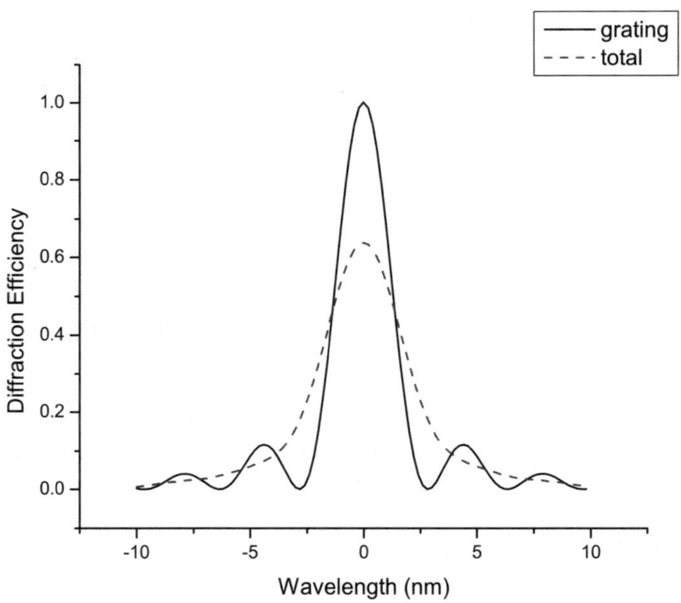

Fig. 11. The spectral response of an EH-based switching device: (*solid line*) the diffraction efficiency of a sinusoidal grating; (*dashed line*) the output power collected by the output collimator

as function of $\delta\lambda$, computed by expression (5). The dashed line presents the power collected by a collimator aligned along the Bragg output angle computed by the convolution of the diffraction efficiency expression (5) with the beam profile. Note that the side-lobes arising from the sinc dependence of η on ξ in the pure grating disappear when the output diversion from the Bragg angle $-\theta_{\mathrm{B}}$ is taken into account. It is clear that the spectral bandwidth of the basic EHD presented in Fig. 11 cannot support dense WDM applications. One approach for narrowing the bandwidth is to enlarge the interaction length traveled by the input beam through the grating. This approach will not yield a substantial improvement unless the device size and the beam diameter are enlarged beyond practicality. An alternative approach is to form a grating with a smaller period. The grating period is determined by the Bragg angle and is limited in the simplistic configuration (Fig. 6) by the critical angle of the crystal. Note that in the device described above, due to Snell refraction at the plane of incidence of the crystal the Bragg angle is $\theta_{\mathrm{B}} = 18°$. This limitation is obviated in the 0–90° configuration presented schematically in Fig. 12 .

Here the angle of incidence into the device is 0° so that Snell refraction does not occur. The grating vector is inclined at 45° to the plane of incidence so that θ_{B}=45°. The spectral response of an EHD built in the 0-90° configuration is presented in Fig. 13.

The grating period is 0.5 μm and the beam diameter is 2 mm. Note that this configuration substantially improves the selectivity. Using this configuration, a selectivity of 100 GHz for DWDM applications is obtainable. It should be emphasized, however, that there is a trade-off between the maximum diffraction efficiency that can be obtained and the selectivity when the grating period is of the order of magnitude of the Debye length in the crystal. Increasing the impurity concentration in the crystal can obviate this trade-off.

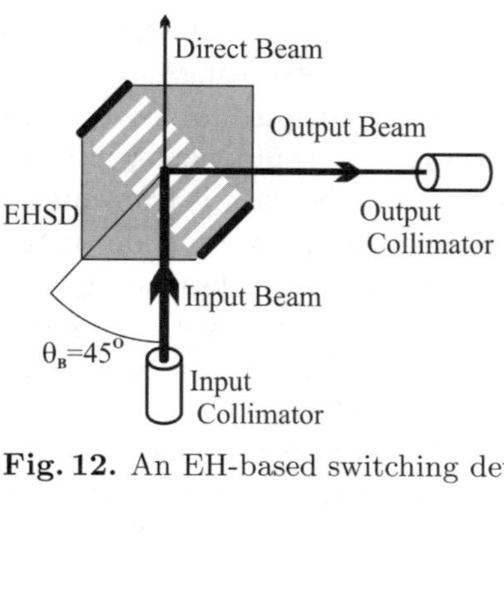

Fig. 12. An EH-based switching device in the 0–90° configuration

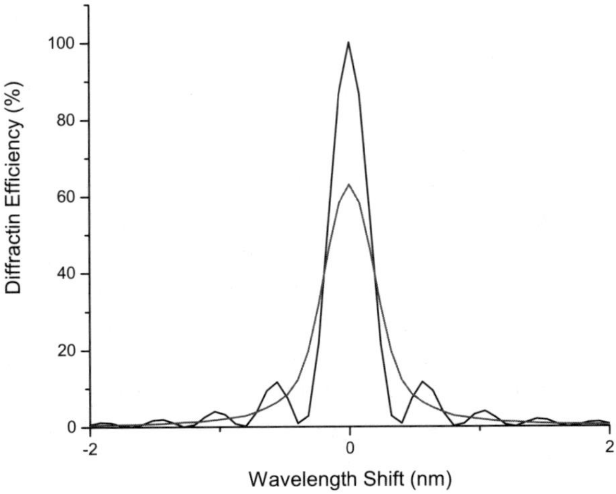

Fig. 13. The spectral response of an EH-based switching device in the 0–90° configuration

4.2.6 Switching Time

We distinguish between two main methods of switching transmitted information across the network: circuit switching and packet switching [25]. The switching-time requirements on the EHD are derived from the switching method employed by the system in which the devices are embedded. Circuit switching systems implement transparent data pipes in which the communication path is intended for long transmissions such as telephone calls. Consequently, set-up and tear-down times of the link and the response time of the system to evolving situations are long, namely, in the milliseconds range. (e.g. the allowed restoration time in a Synchronous Optical Network (SONET) system is 50 ms). The current performance envelope of circuit switching systems is defined under the assumption that the information traffic in the network is primarily voice telephony. As the network will evolve into being more data-oriented, the performance envelope will have to be redefined. It is expected that the required switching time of the basic switching component will be in the microseconds range. In packet switching systems the data is transmitted in packets that are routed towards their destination at every node of the network according to the address stored in their header. Consequently, the switching time of the basic switching component in the packet switching arena should be faster than the length of the packet, namely, in the nanoseconds to submicroseconds range.

The switching time of the EHD is governed by the physical mechanism of the switching operation. Consider expressions (8), (10) and (12) that describe the phenomenology of this mechanism in PR paraelectrics. The application of the electric field to the latent electrohologram first of all produces to induce a polarization in the medium as given by (10). The induced polarization couples with the polarization field induced by the space charge, and through the quadratic electro-optic effect causes a change in the index of refraction. The desired working point of a PR paraelectric EHD implemented in a KLTN crystal is slightly above the Curie temperature. The critical slowing down of the dielectric response in the vicinity of T_c causes the dielectric response (i.e. the time for the applied field to induce the polarization) to be the dominant factor governing the temporal behavior of the switch. The induction of the changes in the index of refraction through the quadratic electro-optic effect is at timescales that are much shorter than the dielectric response of the crystal. Consequently it is expected that the rise time and fall time of the EHD will be in the nanosecond range. Consider Fig. 14 in which the spectral response for an EHD is presented. The diffracted power follows the temporal behavior of the driving electric field.

The rise time demonstrated in Fig. 14 is less than 20 ns (defined as the time needed for the diffracted light to rise from 10% to 90% of its maximum power). Note that as T_c is approached, the slowing down of the dielectric response becomes apparent, causing a trade-off between the response time and the diffraction efficiency manifested in the insertion loss of the device. It is expected that the response time of EH-based switching will not be shorter than a few nanoseconds. Therefore, EH-based switching can become the platform for introducing the WDM technology to the packet/burst switching arena in applications that do not require switching times shorter than a few nanoseconds.

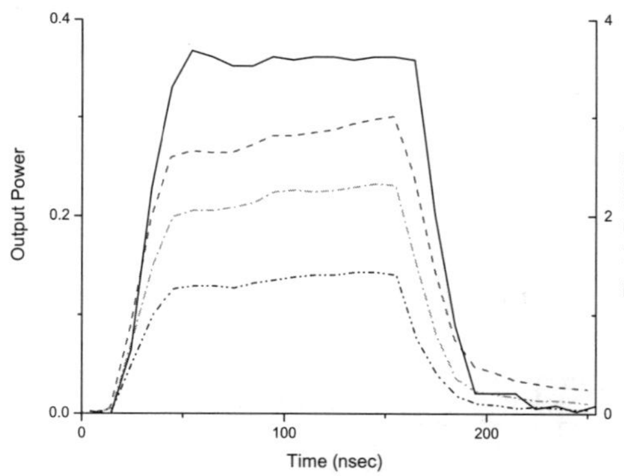

Fig. 14. The temporal response of an EH-based switching device: (–) the applied electric field; the diffracted power at (- -) T = $T_c + 15\,°C$, (-.-) T = $T_c + 20\,°C$ and (-..-) T = $T_c + 30\,°C$

5 Applications of Electroholographic Switching

The EH-based switching device is a generic device, and as such can be the building block in a variety of wavelength-selective switching applications implemented by different configurations. We henceforth present two applications that are based on the EH-based switching device an EH-based dynamic add–drop multiplexer and an EH-based cross-connect.

5.1 The Electroholographic Dynamic Optical Add Drop Multiplexer

The EH-based dynamic optical add drop multiplexer (DOADM) is a straightforward application of electroholography. Consider Fig. 15 in which a possible implementation of an EH-based DOADM unit is illustrated.

The DOADM in Fig. 15 is designed to be installed in a Metro or inter-office ring. The DWDM lightwave entering the unit through the input port is first demultiplexed into the single-wavelength channels. The latter are regrouped by the 'Drop' module into three groups: local 'drop' 1, local 'drop' 2, and an 'express channel' that continues to the output port and onward to the ring. The 'add' module contains two 'add' ports that are coupled to the output port of the unit. In addition the unit contains power management and data monitoring capability attached to the 'drop' module. The specific configuration of the EH-based DOADM illustrated in Fig. 15 involves several design choices:

1. In Fig. 15 a DWDM demultiplexer is used to break the DWDM lightwave into its single-wavelength channels. Alternatively, it can be implemented by a row of EHDs, or in the case of a DWDM lightwave with many wavelength

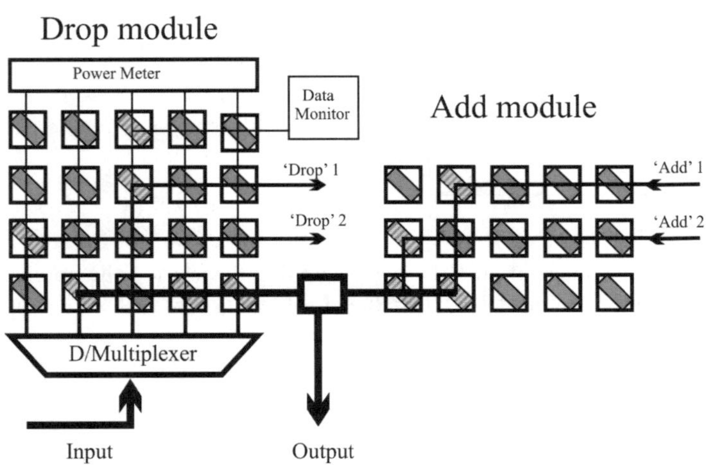

Fig. 15. Electroholographic dynamic optical add-drop multiplexer

channels by a series of interleavers interlacing a series of EH-based 'drop' modules.
2. The number of 'add' and 'drop' channels implemented in the unit illustrated in Fig. 15 is arbitrary and is application-dependent. Moreover, in most applications, each local 'add' port and each local 'drop' port is required to handle one wavelength channel.
3. The incorporation of optical amplifier in the EH-based DOADM is simple since the unit has power management capabilities.

5.2 The Electroholographic Cross Connect

The EH-based cross-connect fully exploits the potential of electroholography as a wavelength-selective switching technology. As such the EH-based cross-connect falls into the system category of "intelligent optical switches" (IOS). The IOS are expected to integrate the isolated DWDM point-to-point links into a truly optical network that is operated in an intelligent and integrated fashion. More specifically the expected benefits of the IOS are improved bandwidth efficiency and scalability, faster provisioning speed, and significant cost, power and footprint savings. A basic cross-connect unit based on EH is illustrated in Fig. 16.

The unit in Fig. 16 is capable of interconnecting four DWDM lightwaves, each carrying N single-wavelength channels. Accordingly, the unit is built of four EHSM, each of which has N wavelength columns, four output rows, a management row, and a power management unit. The respective output rows of the EHSMs are coupled to form the respective output port of the cross-connect. The lightwave emanating from each input fiber is first demultiplexed into its single-wavelength channels, which are then regrouped by the EHSM assigned to this fiber according to their respective destinations. Finally, the four subgroups that are allocated to the same output fiber are coupled together and are transmitted out of the respective output port. Note, that each EHSM is equipped with both power management and data monitoring capabilities that are integrated with the cross-connect operation. The $4 \times N \times 4$ architecture illustrated in Fig. 16 demonstrates the underlying principle of operation of the EH-based cross-connect in its most simplistic form. It does not extract the full potential of EH-based wavelength-selective switching for implementing optical cross-connect architectures. Other design options are possible:

1. The dimensions of the cross-connect, namely, the number of single-wavelength channels per fiber, the number of input ports, and the number of output ports is application dependent.
2. The architecture presented in Fig. 16 is transparent. The single-wavelength DWDM channels propagate through the unit unaffected, assuming that data regeneration is not necessary. (Note that the latter is inherently electronic.) The EH-based cross-connect enables the integration of transparent

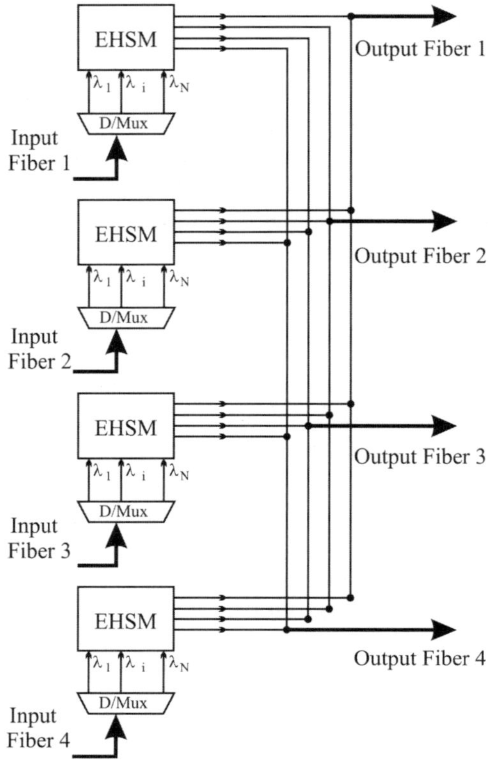

Fig. 16. Electroholographic cross-connect

and opaque switching in the same unit on a cost-effective basis. This is accomplished by assigning an input fiber and an output fiber to the task. The output fiber is connected to the input of an Opto-electro-optic (OEO) unit. The latter receives the wavelength channels in the form of a DWDM lightwave, demultiplexes them and performs the regeneration. It then regroups these channels into the input channel, transmits them back to the cross-connect, where they are redistributed to their final destinations. Thus, the EH-based cross-connect enables to limit the complex and expensive OEO operations only to the channels for which it is required.

3. Note that in the architecture illustrated in Fig. 16 two channels of the same wavelength originating from two distinct input fibers may be allocated to the same output fiber. In this respect this architecture is 'blocking'. The system manager can avoid this situation when allocating the single-wavelength channels to the output ports. It can also be avoided by the incorporation of wavelength conversion in a similar way to the incorporation of the regeneration operation as described above. Here as well the incorporation of wavelength conversion is cost effective as it is limited to the level that is statistically anticipated by the system designer.

4. The incorporation of optical amplifiers in the EH-based cross connect is simple since each of the EHSM has power management capabilities.

6 Conclusions

WDM and in particular dense WDM have dramatically improved the capability of the 'long haul' segments of the networks to sustain a tremendous growth in data traffic. It is now well established as the reigning technology for point-to-point links in the capable of sustaining hundreds of Gigabits per seconds in a single fiber. However, the use of DWDM technology in the optical networks is limited to 'long haul' segments where it provides extremely efficient and cost effective point-to-point links. EH can supply the platform for viable wavelength-selective switching technology that will enable to intelligently integrate the isolated disparate WDM point-to-point links into cohesive agile networks.

References

1. A. J. Agranat: Electroholographic artificial neural networks, Physica A **200**, 608-612 (1993)
2. A. J. Agranat: The Concept of Electroholography, Its Implementation in KLTN crystals, in S. C. Esener (Ed.), *Proc. of OSA Topical Meeting on Spatial Light Modulators* (Lake Tahoe, Nevada 1997)
3. M. Balberg, M. Razvag, S. Vidro, E. Refaelli, A. J. Agranat: Electroholographic neurons implemented on potassium lithium tantalate niobate crystals, Opt. Lett. **21**, 1544–1546 (1996)
4. B. Pesach, G. Bartal, E. Refaeli, A. J. Agranat: Free space optical cross-connect switch by use of electroholography, Appl. Opt. **39**, 746 (2000)
5. H. Kogelnik: Coupled wave theory for thick hologram gratings, Bell Syst. Tech. J. **48**, 2909 (1969)
6. L. Solymar, D. J. Cooke: *Volume Holography and Volume Gratings* (Academic, London 1981)
7. A. Yariv: Two-beam coupling and phase conjugation in photorefractive media, *Optical Electronics* (Saunders College Publ., Philadelphia 1991) pp.637–668
8. D. B. Nolte, Photorefractive transport and multi-wave mixing, in D. B. Nolte (Ed.), *Photorefractive Effects and Materials* (Kluwer, Dordrecht 1995) pp. 1–66
9. L. Solymar, D. J. Webb, A. Grunnet-Jepsen: *The Physics and Applications of Photorefractive Materials* (Clarendon, Oxford 1996)
10. P. Yeh: *Introduction to Photorefractive Nonlinear Optics* (Wiley, New York 1993)
11. J. E. Geusic, S. K. Kurtz, L. G. Van Uitret, S. H. Wemple: Electro-optic properties of some ABO3 perovskites in the paraelectric phase, Appl. Phys. Lett. **4**, 141 (1964)
12. A. J. Agranat, V. Leyva, A. Yariv: Voltage-controlled photorefractive effect in paraelectric $KTa_{1-x}Nb_xO^3$: Cu, V, Opt. Lett. **14**, 1017-1019 (1989)

13. J. P. Wilde, L. Hesselink: Electric field controlled diffraction in photorefractive strontium barium niobate, Opt. Lett. **17**, 853-855 (1992)
14. A. J. Agranat, R. Hofmeister, A. Yariv: Characterization of a new photorefractive material: $K_{1-y}Li_yTa_{1-x}Nb_xO_3$, Opt. Lett. **17**, 713-715 (1992)
15. C. H. Perry, R. R. Hayes, N. E. Tornberg, Ferroelectric soft modes and anti resonance behavior, in M. Balkanski (Ed.), *KTN Mixed Crystals*, Proc. Int. Conf. on Light Scattering in Solids (Wiley, New York 1975) 812-815
16. A. J. Agranat, R. Hofmeister, A. Yariv: Potassium lithium tantalate niobate photorefractive crystals, U. S. Patent **5, 785, 898**, (1998)
17. A. J. Agranat, R. Hofmeister, A. Yariv: KLTN A new photorefractive material for holographic data storage and optical computing applications, U. S. Patent **5, 614, 129** (1997)
18. R. Hofmeister, S. Yagi, A. Yariv, A. J. Agranat: Growth and characterization of KLTN:Cu, V photorefractive crystals, J. Cryst. Growth **131**, 486 (1993)
19. A. J. Agranat, L. Secundo, N. Golshani, M. Razvag: Wavelength-selective photonic switching in paraelectric KLTN, Opt. Mater. **18**, 195 (2001)
20. G. Bitton, Y. Feldman, A. J. Agranat: Dielectric properties of photorefractive KLTN crystals, Ferroelectrics **239**, 213 (2000)
21. F.H. Mok, G.W. Burr, D. Psaltis: Opt. Lett. **21**, 896-898 (1996)
22. B. Pessach, E. Refaeli, A. J. Agranat: Investigation of the holographic storage capacity of paraelectric KLTN, Opt. Lett. **23**, 642-644 (1998)
23. M. Puterkovky: The photorefractive sensitivity and holographic storage capacity in KLTN and KNTN crystals, M. Sc. Thesis, The Hebrew University of Jerusalem (2001)
24. K. Blotekjaer: Limitations on holographic storage capacity of photochromic and photorefractive media, Appl. Opt. **18**, 57 (1979)
25. L. Kazovsky, S. Benedetto, A. Willner: *Optical Fiber Communication Systems* (Artech House, Boston 1996)

1550 nm Volume Holographic Devices for Optical Communication Networks

Pierpaolo Boffi, Maria C. Ubaldi, Davide Piccinin, and Mario Martinelli

CoreCom, Via Ampère-20131 Milano, Italy
boffi@corecom.it

Abstract. The present contribution aims to show the feasibility of $LiNbO_3$:Fe volume holography (VH)-based devices for optical fiber communication networks. The VH technique offers a valid alternative to the existing approaches in the building of multiplexers/demultiplexers and databases for individual wavelengths inside an optical wavelength division multiplexing (WDM) system. The use of angle multiplexing jointly with the two-lambda method and the thermal post-fixing technique allow us to achieve efficient and long-lifetime operation in the near-infrared spectral range. Optical fiber communications are rapidly growing in traffic owing to many new important services such as mobile telephony and Internet connections. Increasing capacity demand calls for more transmission bandwidth and higher bit rates. In order to exploit the entire spectrum of the low-loss regions of the fiber attenuation window, Wavelength Division Multiplexing (WDM) transmission mode is today in common use. WDM technology combines multiple optical signals into a single fiber by transmitting each signal on a different wavelength (as happens in the radio spectrum). This means that telecom carriers can multiply the capacity of their fibers without the expensive investment of laying more fiber underground and undersea.

1 Introduction

In order to provide high capacity the WDM system has to allow broader optical bandwidth (usually the whole fiber third transmission window, around 1550 nm), higher bit-rate per channel (up to 40 Gbit/ s) and minimum channel spacing (say from 100 to 25 GHz). In this context the electronics presents the real bit-rate bottleneck inside the network: it is important to partially replace it with optical technologies, by maintaining the optical nature of the signals, in order to build a future all-optical network. Moreover, WDM channels do not ideally interact each other and this allows us to add as many channels as required, where each one can operate at a high data rate. In practice, optical crosstalk occurs and the number of channels is therefore limited by the performance of each individual component and device in the system.

All these requirements challenge component designers to devise novel solutions within the given technical and cost constraints. The next optical communication network will need novel all-optical components and devices which will be able to manage both simultaneously and in real time all the channels at the different wavelengths propagating in the WDM system. Volume

P. Boffi, P. Piccinin, M. C. Ubaldi (Eds.): Infrared Holography for Optical Communications,
Topics Appl. Phys. **86**, 157–178 (2003)

holography (VH) could be the ideal candidate technology providing for the implementation of such devices.

The first researcher to use VH was *Denisyuk* in 1962 in order to improve the fidelity of holographic reconstructions [1]. With reference to planar holography in thin supports, the main VH feature is the presence of the additional thickness size, which allows the recording of multiple holograms superimposed into the same volume of photorefractive material. The parallel fast access to multiple stored information exploits the massive parallelism which is typical of optics, without any optoelectronic and electro-optic conversions.

The most common methods for superimposing multiple holograms include angle multiplexing and wavelength multiplexing [2,3,4]. It is possible to transfer the powerful advantages of volume holography in terms of high readout selectivity in angle or in wavelength to WDM applications by associating each recorded hologram to a specific readout wavelength corresponding to a WDM channel.

Unfortunately there is a shortage of photorefractive materials with good holographic sensitivity in the near-infrared (NIR) spectrum, i.e. the optical communication transmission range. Standard photorefractive materials demonstrate optical sensitivity peaks in the visible spectral range. It is now still hard to find a material which can offer an adequate sensitivity in the third spectral window of optical communications. Some specially doped II–VI semiconductors, for example vanadium-doped cadmium telluride (CdTe:V) can perform a certain amount of photorefractive efficiency at 1550 nm (typically less than 10^{-3} [5,6]), but the experimental value is too low for recording directly at 1550 nm holographic gratings that are useful for fiber communication optical devices.

However, the so-called two-lambda method [7] offers the opportunity of both recording holograms in standard photorefractive materials by means of light at maximum sensitivity wavelengths and reading such holograms at NIR wavelengths.

In this chapter, we will demonstrate the feasibility of VH-based optical devices fabricated in standard iron-doped lithium niobate ($LiNbO_3$:Fe) operating in the third optical communication window. In particular we will consider the implementation of both a WDM demultiplexer and a holographic digital database readable by WDM beams. The long-lifeterm stability of such devices is also taken into account in order to permit the fabrication of new commercial communication network devices.

2 Building the Optical Communication Network

Figure 1 shows a typical optical WDM system layout. Just by designing new optical components and devices to improve the system capacity, it is possible to meet the demands of next-generation optical networks. In particular we analyze some essential network devices in order to identify their performance

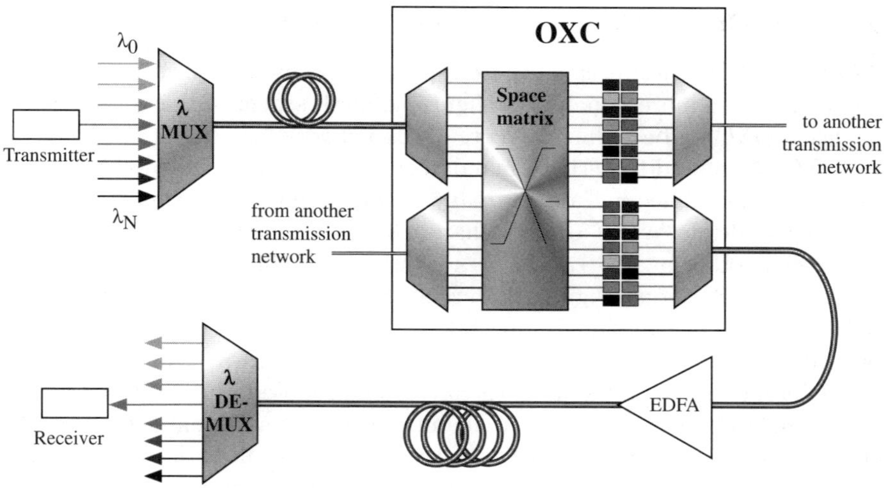

Fig. 1. Typical optical WDM system. The elements drawn in grey standing represent optical devices which can be designed by using holographic technology

and limitations and to understand their capability for exploiting innovative solutions based on VH technology.

2.1 The Optical Wavelength Multiplexer/Demultiplexer

The optical mux/demux combines/separates out WDM channels at different wavelengths. For demultiplexing a minimum of optical crosstalk the demultiplexed channels has to be guaranteed. With a large wavelength spacing between two neighboring channels the above requirement appears relatively easy to satisfy, especially for low data rates. However, for a reduced wavelength spacing, the device design has to be much more sophisticated, requiring a high rejection over a wide stopband, which may be difficult to achieve with standard period-response filter designs. The adopted choice is now the use of arrayed waveguide gratings (AWGs) or fiber Bragg gratings (FBGs).

AWGs [8], being integrated, have an inherent size advantage. By separating the input signal into n planar waveguides, each designed with a different path length, and recombining them in a $n \times n$ coupler, the WDM channels become spatially separated through interference and diffraction effects. AWGs imply high insertion losses and significant polarization-dependent losses.

FBGs [9] present instead a very sharp filtering function, allowing channel spacing up to 50 GHz. Demultiplexing is achieved by a cascade of FBGs, but optical circulators or interferometer schemes are necessary in order to extract the reflected demultiplexed channel without too many losses.

The wavelength-multiplexing operation can be implemented by means of the same solutions, utilized in the inverse way.

2.2 The Optical Cross-Connectand the Switching Fabric

As optical WDM networks move from point-to-point configurations and ring deployments to more dynamic configurable mesh schemes, the optical cross-connect (OXC) is becoming the core of the network.

OXC is required to perform mandatory functions such as switching, routing and cross-connecting among many input/output fiber trunks, each one carrying multiple WDM optical channels. These functions are most often performed today by the equipments related to client electronic layers (for example Synchronous Digital Hierarchy (SDH)). In order to maintain the optical nature of the signals, the OXC replaces electronic operations with optical devices such as the space switching matrix and wavelength converter.

Different OXC architectures [10] have been proposed: the space matrix can be realized by means of single-switch units based on traditional optical effects, such as electro-optical and thermo-optical ones. Till now the implemented space switches cannot be reliably manufactured at densities over 8×8 or 16×16. The front-runner among optical switching solutions is today constituted by microelectro mechanical systems (MEMS) [11], which provide a matrix of up to 256×256 elements on a very small chip.

An innovative free-space switching matrix based on electroholographic technology is now commercially available too [12]. In this solution the diffraction process of the stored volume holograms is controlled by means of an externally applied electric field. It appears very attractive thanks to its inherent very low crosstalk and the ease of three-dimensional routing.

Moreover, in the case of packet-oriented data networks such as Internet Protocol (IP) or Asynchronous Transfer Mode (ATM), directly traveling on the WDM transport layer, routing is performed by the switching fabric enriched by more functionalities in order to process these packets. For example, the packet has to be identified and rerouted. The packet header is modified with new information related to the next address, where the packet has to be routed. This information is contained in suitable databases, often called look-up tables. In the case of a WDM network, these tables can be interrogated through the wavelength and contain information associated with the WDM channel identifier. Digital data management is currently performed by optoelectronics: in the case of a very high transmission bit rate, in order to avoid an explosion in complexity and costs, new photonic solutions are required, especially for information storage and retrieval.

3 Volume Holography for 1550 nm Optical Device Implementation

In photorefractive crystals, VH thanks to its inherent selectivity properties stands out as having much promise for optical devices for future fiber communication networks. In this paragraph we report some theoretical considerations related to multiplexing techniques and to the so-called two-lambda method, which is useful for reading holograms recorded in the visible spectral range by means of beams at different wavelengths (in particular in the NIR range). We shall not focus our explanation on the principles of the photorefractive effect, referring to the literature [13]. Only thermal fixing considerations [14] applied to recorded multiple holograms will be discussed in detail.

3.1 Recording and Readout of Multiple Holograms

The angle-multiplexing technique [2] identifies the recording of superimposed holograms by either varying the angle of incidence the recording beams or the grating direction inside the crystal. In the latter situation the grating vectors associated with the holograms are equal in modulus, but different in orientation (Fig. 2). In Fig. 2a the crystal rotation has been indicated as an equivalent rigid rotation $\Delta\,\theta_{\mathrm{mux}}$ of the recording beams for simplicity. The vectors associated with the optical beams and the grating can be visualized easily in a graphical representation called k-space (Fig. 2b). Considering to operate at the wavelength λ, the set of possible beams spans a sphere of radius $2\pi/\lambda$. For simplicity, a circular 2-D slice through the sphere is shown. The recorded grating is represented as the vector that joins the two beam wave vectors. When the readout beam incidence angle is identical to the recording one, the Bragg-matched condition is obtained (the triangle constituted by the three vectors in k-space is closed). The readout beam is diffracted only by the proper hologram. All the other stored holograms are suppressed by Bragg mismatch. The diffraction efficiency of each stored hologram is a strong function of the readout condition in terms of angle: here the angle spacing between the maximum diffraction efficiency and the first null is usually called selectivity $\Delta\,\theta_{\mathrm{B}}$. A high selectivity (small $\Delta\,\theta_{\mathrm{B}}$) is achieved by increasing the grating depth L, due to the large thickness of the photorefractive crystal. Wavelength-multiplexed [3] readout is possible by leaving unchanged the readout beam incidence angle while varying its wavelength (Fig. 3).

Any number of holograms can be recorded inside the same crystal. The main problem concerning multiple hologram recording is the crosstalk, defined as the ratio between the maximum intensity diffracted from the read hologram and the intensity diffracted simultaneously in the same direction by adjacent holograms. In order to achieve low crosstalk, the optimal condition is to choose the angular spacing $\Delta\,\theta_{\mathrm{mux}}$ larger than 3–4 times $\Delta\,\theta_{\mathrm{B}}$ for neighboring gratings. Finally, the recorded multiple holograms have to have the same diffraction efficiency. Unfortunately, at the maximum sensitivity wavelengths for the photorefractive crystal, the recording beams required to store a new hologram tend to erase the previously recorded gratings. It is necessary to know the erasure time constant in order to carefully schedule the recording process with a set of decreasing exposure times [15]. Initial holograms have to be recorded with a large diffraction efficiency and erased

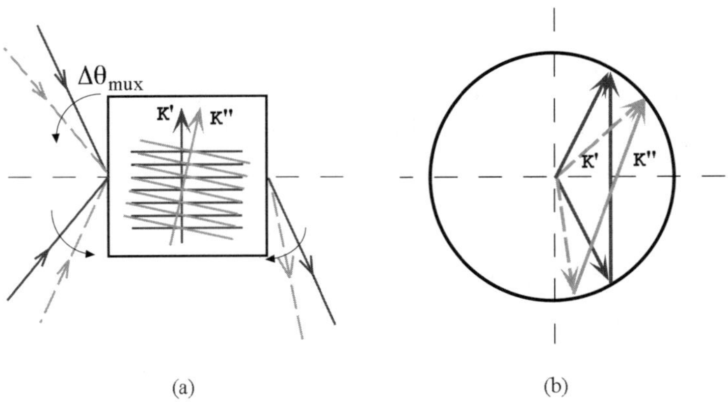

Fig. 2. Angle-multiplexing technique: representation (**a**) in real space and (**b**) in k-space

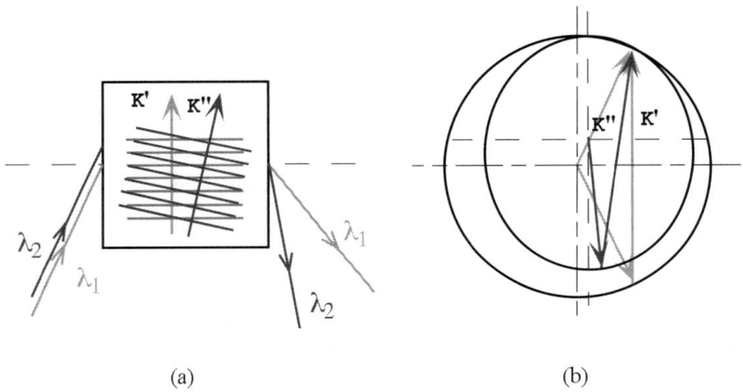

Fig. 3. Wavelength-multiplexed readout: representation (**a**) in real space and (**b**) in k-space

by subsequent exposures at the same diffraction efficiency as the final short exposure.

3.2 Two-Lambda Method

The two-lambda method has been widely studied in the literature in the visible range [7,16,17], the primary aim being the prevention of optical erasure of the recorded holograms. It is possible to exploit this technique also in the NIR spectral range in order to transfer the advantages of VH to the field of optical fiber communications. A hologram is recorded at the writing wavelength λ_w with an incidence angle θ_1 for the object beam and θ_2 for the reference one (with respect to the normal to the crystal incidence face). The recorded hologram can be reconstructed at a different wavelength λ_r by introducing a readout beam at an angle ϕ_1 tilted with respect to the writing reference beam. The Bragg law implies that [7]:

$$\frac{\lambda_\mathrm{w}}{\sin(\frac{\theta_1+\theta_2}{2})} = \frac{\lambda_\mathrm{r}}{\sin(\frac{\phi_1+\phi_2}{2})}, \tag{1}$$

$$\theta_1 - \theta_2 = \phi_1 - \phi_2, \tag{2}$$

where ϕ_2 defines the direction of the diffracted beam at readout wavelength. The same condition can be easily shown in the k-space diagram: the satisfaction of the Bragg law is graphically represented by the closing of the triangle formed by the wave vectors of the incident ($\boldsymbol{k}^\mathrm{i}_{\mathrm{w,r}}$) and diffracted ($\boldsymbol{k}^\mathrm{d}_{\mathrm{w,r}}$) beams, and the grating wave vector $\boldsymbol{K}$ (Fig. 4b).

It is clear that a change in the readout wavelength must be compensated by a variation of the incidence angle of the readout beam, in order to satisfy the Bragg law. While we can easily Bragg-match a single grating, when a 2-D hologram consisting of many plane waves components is recorded, it is impossible to match the whole spectrum with a single plane readout wave. Recording a 2-D image can be represented in k-space by a cone of vectors that interferes with the reference beam to record a cone of grating vectors (Fig. 5a). When we want to reconstruct the 2-D image at a different wavelength, only the gratings lying on the circle of intersection between the two k-spheres are actually Bragg-matched. As a consequence, only an arc of the image cone can be reconstructed in output with good diffraction efficiency. The effects on the reconstructed image are dependent on whether we record in the Fourier plane or the image plane. By recording in the Fourier plane, as a result of the retrieval process just a strip of the image (in the direction perpendicular to the incidence plane of the recording beams) is reconstructed (Fig. 5b,c [18]).

By image plane recording (i.e. in place of the input image we use its Fourier transform), only a band of the image frequency spectrum can be retrieved: in particular, if the readout beam is Bragg-matched with the DC component of the image, the reconstruction at λ_r will introduce a low-pass filtering effect on the original image.

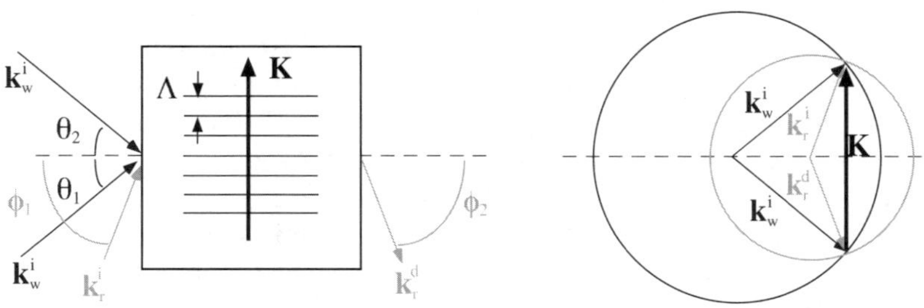

Fig. 4. Wave vector representation of two-lambda method: (**a**) inside the crystal and (**b**) by means of a k-space diagram

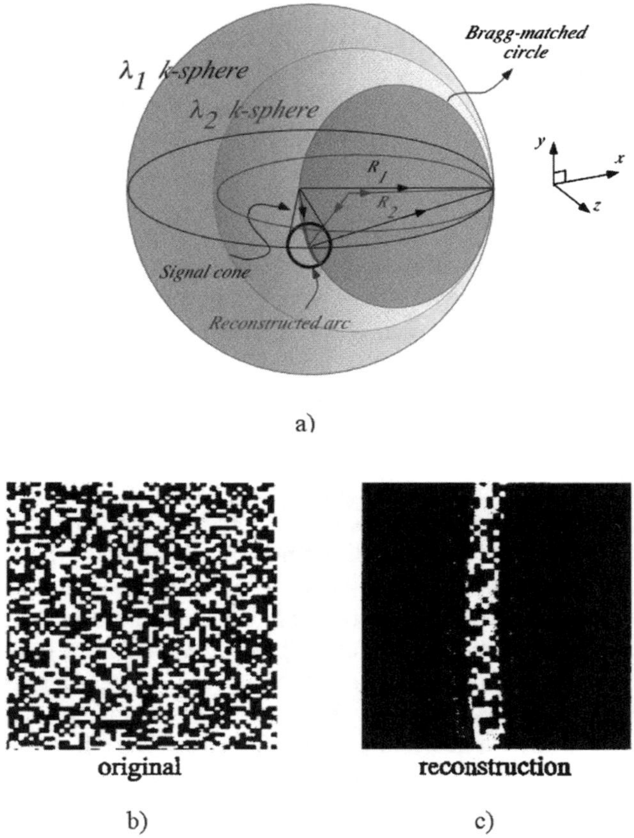

Fig. 5. (**a**) k-sphere diagram for the two-lambda method geometry with a 2-D recorded signal, (**b**) input image at λ_w and (**c**) its reconstruction at λ_r

3.3 Long-Term Lifetime by Thermal Fixing

The volatility of stored holograms in photorefractive materials is the main drawback of the holographic technique for fabricating commercial devices. Photorefractive grating fixing is necessary. Thermal fixing seems to be the most suitable and easy method to use (thermal fixing of 10^4 angle-multiplexed holograms at 514 nm has been demonstrated [19]). It is typically achieved by means of a post-fixing technique. The hologram is recorded at room temperature; then the photorefractive crystal is h eated at a temperature between 100 °C and 180 °C and subsequently cooled down again at room temperature (RT) (low-high-low cycle). The whole process dynamic is shown in Fig. 6.

In phase (I) the recorded electronic grating is heated to increase the protonic mobility; this leads to a so-called compensated hologram which is associated with the net space-charge field $E_1^{(1)}$. During the development phase (II) the grating is illuminated by incoherent or coherent light (in the latter case

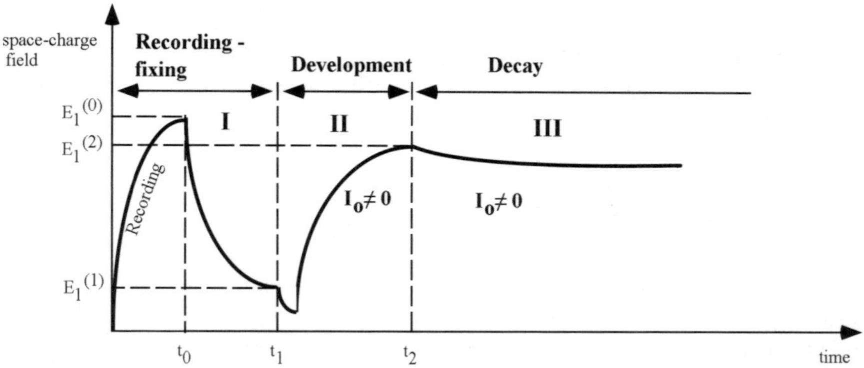

Fig. 6. Space-charge field behavior during holographic recording, thermal fixing, development and decay

incident far from the Bragg angle) comparable in intensity (I_0) to the recording beams; the trapped electrons are redistributed until the ionic gradient is completely revealed, leading to a quasi-stable field $E_1^{(2)}$. In phase (III) the decay process under illumination is essentially due to ionic transport.

After this brief description of the fundamental steps related to the ionic grating formation, we provide some physical details. Protons become mobile at high temperatures and tend to compensate the electric field responsible for the refractive index grating due to the electro-optic effect: the formation of a non-uniform ionic distribution gradually reduces the diffraction efficiency. After cooling down to RT the protons become immobile; during the developing stage the electrons are wrenched from the impurity traps and can freely move inside the conduction band. The positive charges are no longer balanced and an ionic replica of the refractive-index modulation is obtained. At the end of the process a determined fraction of original diffraction efficiency is reconstructed and in principle the ionic hologram is non-photosensitive. As a matter of fact the ionic mobility is negligible but not zero at RT; their conductivity is responsible for a decrease in diffraction efficiency over an extremely long time, typically tens of years. The two most important features characterizing the performance of a thermal fixing process are surely the ionic hologram lifetime τ and the reconstructed diffraction efficiency η_{fixing}. The hologram lifetime constant is defined as

$$\frac{1}{\tau} = \Gamma_{\text{s}} = D_{\text{H}} K^2 \left(\frac{H_0}{N_{\text{t}}} + 1 \right) , \tag{3}$$

where D_{H} is the protonic diffusion constant, H_0 the protonic concentration and $K = 2\pi/\Lambda$ with Λ the grating period. N_{t} is related to the doping ion concentrations; for iron doping $N_{\text{t}} = [\text{Fe}^{2+}][\text{Fe}^{3+}]/([\text{Fe}^{2+}] + [\text{Fe}^{3+}])(\approx \text{Fe}^{2+}$ for oxidized samples). The highest hologram stability corresponds to the maximum value of τ, i.e. $\tau = (D_{\text{H}} K^2)^{-1}$, which occurs for $H_0/N_{\text{t}} \ll 1$; thus the lifetime can be increased by both increasing the grating spatial frequency and reducing the hydrogen content of the crystal. The fixing efficiency is the portion of the strength of the original hologram attained after the revealing process and is given by

$$\eta_{\text{fixing}} = \frac{E_1{}^{(2)}}{E_1{}^{(0)}} = \frac{(\frac{(N_\text{d}-N_\text{a})E_{0\text{ph}}}{N_\text{d}})^2 + {E_\text{D}}^2}{(\frac{(N_\text{d}-N_\text{a})E_{0\text{ph}}}{N_\text{d}})^2 + (E_\text{D}+E_\text{q})^2}\,, \tag{4}$$

where $E_{0\text{ph}}$ is the photovoltaic field, E_D is the diffusion field, E_q is the saturation space-charge field, and N_d and N_a are the donor and acceptor concentrations,respectively. It should be noted how the saturation space-charge field E_q affects η_{fixing}. In fact, if this field is small, the efficiency can approach unity, reaching the value achieved after the recording process; this situation occurs for both low dopant concentration and short grating periods.

4 VH-Based Devices for WDM Applications

Nowadays research is strongly stimulated to explore the capabilities of optics, not only for signal transmission, but also for signal processing. The implementation of all-optical techniques such as VH appears very attractive for WDM applications, where many channels at different wavelengths travel inside the same fiber. VH-based devices can manage the different WDM channels simultaneously, in real time and in a passive way, without slow and inefficient optoelectronic conversion, by associating each one to the proper recorded hologram. Here we present the design, development and preliminary testing of two particular devices, previously described in Sect. 2 as key components of the optical communication network. In both the implementations, superimposed holograms are recorded in a 0°-cut lithium niobate crystal 0.015 mol% iron-doped ($LiNbO_3$:Fe). The use of such a standard photorefractive material guarantees large recording capacity and high angular selectivity, associated with a long storage lifetime and very high diffraction efficiency. The operation of the implemented devices undder severe environmental conditions and their estimated long-term reliability are also discussed. Further improvements in the device design are also considered in order to deal with future high-dense WDM (HDWDM) needs.

4.1 Feasibility of an Optical Wavelength Demultiplexer

As described in Sect. 2.1, the optical wavelength demultiplexer separates out WDM channels. Its implementation is made possible by means of a crystal containing previously recorded diffraction gratings with a suitable angular separation [20]. Each WDM channel at a different wavelength is diffracted only by its own Bragg-matched grating and emerges from the crystal along a different direction, satisfying equations (1) and (2). Hence the WDM channels, which are collinear in input, are demultiplexed in output into angularly separated directions (Fig. 7), so that they can be subsequently coupled into new optical fibers in order to continue their propagation in the network.

The inherent parallelism of optics is exploited in the simultaneous processing of all the WDM signals. The operation is performed in a passive way, totally transparent to the transmission bit rate of the channels to be multiplexed. The same device scheme, but operating in the inverse way, can be utilized in order to perform wavelength multiplexing.

Experimentally, in order to demonstrate the feasibility of a VH-based demultiplexer, a series of four gratings has been recorded into a $LiNbO_3$:Fe crystal, $1 \times 1 \times 1\,cm^3$ in size. The analysis of the behavior of diffraction efficiency vs. both readout angle and wavelength in the third optical communication window spectrum, i.e. around 1550 nm, is necessary to validate the theoretical principles discussed in Sects. 3.1 and 3.2. Angle-multiplexed holograms are recorded at 488 nm, with a crossing angle of 30° in air and an angular separation $\Delta\,\theta_{mux}$ between adjacent gratings of 0.055°; the resulting grating wave vector is parallel to the crystal optical axis $\vec{z}$. In order to get higher diffraction efficiency in lower recording times, extraordinary polarization has been preferred rather than the ordinary one: the recording process at 488 nm in the $LiNbO_3$ crystal allows us to achieve up to $85-90\%$ efficiency for visible light. Long exposure times are required in order to achieve the very high local refractive-index modulation necessary for an efficient readout process at 1550 nm: in fact an increase in $\Delta\,n$ must compensate a reduction of roughly a factor of 3 in the readout wavelength, the diffraction efficiency being expressed as

$$\eta = \sin\left(\frac{\pi\Delta\,nL}{\lambda\cos(\theta_B)}\right)^2 , \tag{5}$$

where $\Delta\,n$ is the photoinduced variation of the refractive index, L is the effective grating depth and θ_B is the Bragg angle. The performed recording exposure times are decreasing, a suitable time scheduling having been chosen in order to achieve a uniform diffraction efficiency for all the diffracted

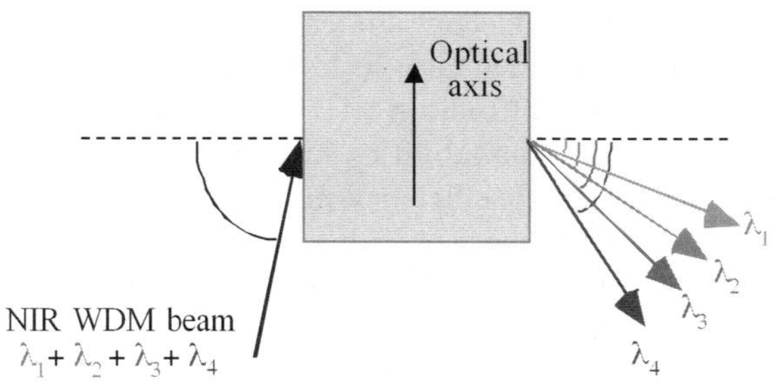

Fig. 7. Schematic of a holographic WDM demultiplexer

Table 1. Parameters related to (*upper*) angle diversity NIR readout and (*lower*) wavelength diversity NIR readout of the four superimposed holograms

♯ hologram	Incidence angle ϕ_1	λ_r(nm)	Diffracted angle ϕ_2
1	55°	1544.5	55°
2	54.91°	1544.5	55.09°
3	54.82°	1544.5	55.18°
4	54.73°	1544.5	55.27°

♯ hologram	Incidence angle ϕ_1	λ_r(nm)	Diffracted angle ϕ_2
1	55°	1544	55°
2	55°	1545.6	55.17°
3	55°	1547.2	55.34°
4	55°	1548.8	55.51°

beams. The experimentally measured parameters for the superimposed holograms are reported in Table 1. The upper part of Table 1 is related to the case of angle-diversity readout, but for the two-lambda condition. The readout λ_r is here fixed at a NIR wavelength, higher than the recording λ_w. The incidence angle ϕ_1 is adjusted in order to satisfy the Bragg law. In the case of the wavelength-diversity readout (lower part of Table 1) the incidence angle ϕ_1 is fixed at 55°, while the readout wavelength λ_r is varied in the third window spectrum. The spacing between the wavelengths diffracted by adjacent gratings is about 1.6 nm and the diffraction angular spacing is 0.17°. Figure 8 shows the diffraction efficiency as a function of the readout wavelength. Four peaks are visible, each one corresponding to a single diffracted hologram. Wavelength demultiplexing is thus achieved. The recording angular separation $\Delta\theta_{\mathrm{mux}}$ of 0.055° has been chosen in order to allow during readout the separation of WDM channels spaced by 1.6 nm, a well-known International Telecommunications Union-Telecommunication (ITU-T) standard wavelength spacing corresponding to 200 GHz in frequency.

The slight difference in the measured diffraction efficiency within the set of stored holograms was caused by laser instabilities in our set-up during the recording process. The measured average efficiency in NIR is about 11%: this value can be increased by adopting higher recording times. The average value of the crosstalk, that is to say how much the single hologram efficiency is affected by contributions due to the neighboring gratings, is around −10 dB, which is rather high for the real employment of such a device as a WDM

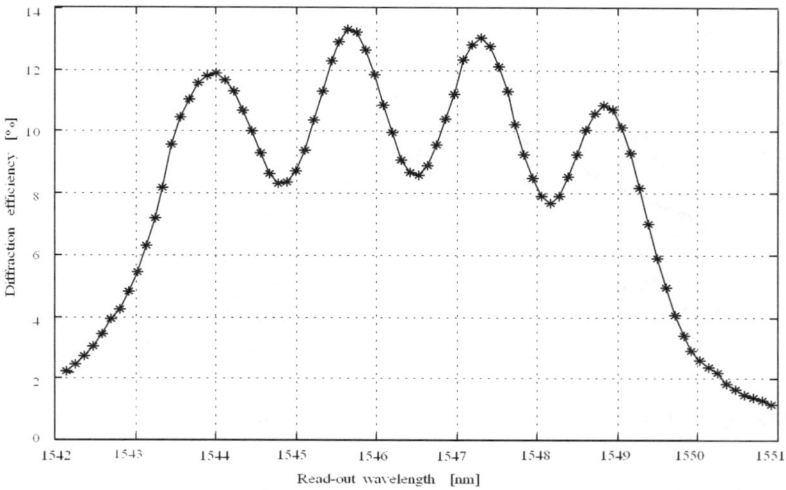

Fig. 8. Wavelength diversity NIR readout of four holograms recorded at 488 nm with 0.055° angular separation. Demultiplexing of 200 GHz-spaced WDM channels is achieved

demux. However, this performance can be strongly improved by adopting different recording and readout geometries, as we will discuss in Sect. 6.

4.2 Feasibility of a WDM-Readable Digital Database

If the VH crystal previously described in Sect. 4.1 contains not simple superimposed gratings, but holograms diffracting a specific digital image, a database readable by means of WDM beams can be implemented [21]. In hologram recording the object beam is no more a plane wave, but a spatially modulated wave carrying n-bit information. From the input WDM optical beams the device generates output beams with different digital images, each one corresponding to a specific wavelength (Fig. 9). In a WDM network, this device can perform the storage functions described in Sect. 2.2

The reconstructed information appears as a 2-D image: it is clear that this space-coded information has to be converted into a temporal sequence of digital data at a suitable transmission bit rate, ready to be transmitted into a fiber link. This conversion can be performed by means of a system of optical delay lines and couplers. The main advantages related to the employment of VH in the realization of such a memory lie both in very fast access and increased storage capacity with respect to traditional media. For 3-D VH memories, a storage capacity of up to 10 GByte/ cm^3 and database access higher than 1 Gbit/ s have been demonstrated [22].

Experimentally, we record inside the crystal 8-bit words space-coded along a direction perpendicular to the incidence plane of the recording beams. In our implementation, the choice of this particular image (encoded along the

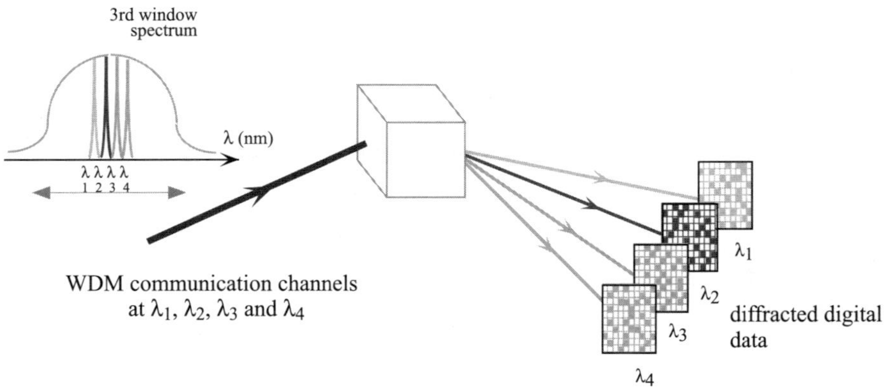

Fig. 9. Digital VH-based database readable by WDM signals

above described direction) prevents any information loss due to retrieval at a wavelength different from the recording one, as described in Sect. 3.2. In fact, while the bit "0" corresponds to the absence of a photorefractive grating inside the crystal, the bit "1" is recorded as a simple interference between two plane waves, thus allowing complete reconstruction of the stored information. In order to implement a holographic device able to perform optical database functionality, multiple byte holograms have been superimposed by the angle multiplexing technique into the same volume of photorefractive material. During recording, suitable transparencies have been interposed along the beam optical path so that the byte image to be stored is obtained. Four digital bytes have been recorded. The recording process is analogous to the one described in the previous paragraph. The angle of incidence θ_1 $(= \theta_2)$ of the 488 nm beam angle is 15° and the angular separation $\Delta\,\theta_{\mathrm{mux}}$ between adjacent holograms is now reduced to 0.03° in order to achieve a wavelength spacing in WDM readout beams of 0.8 nm, corresponding to the ITU-T standard spacing of 100 GHz. By using wavelength-diversity readout, each input WDM beam at a different wavelength is simultaneously diffracted just by its own Bragg-matched hologram, reconstructing in the output the associated digital byte.

Table 2 reports the experimental parameters of the implemented device. Figure 10 shows the infrared camera acquisition of the reconstructed byte 01010101 at 1546.4 nm, where the crystal and the diffracted image are clearly visible. The operation of the 100 GHz WDM readable database is shown in Fig. 11, where the four stored digital space-coded bytes (upper) and the acquisition of the diffracted images (lower) each reconstructed by the related WDM beam, are visible.

Fig. 10. Picture of the byte 01010101 reconstructed in the NIR

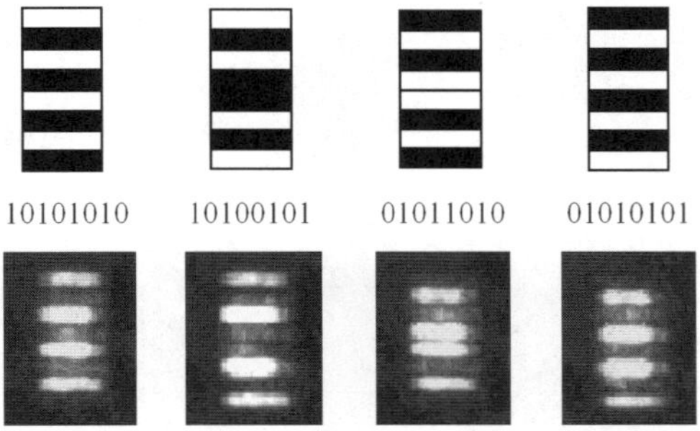

Fig. 11. *Upper*: spatial representation of the recorded bytes. *Lower*: infrared camera acquisition of the diffracted bytes

Table 2. Readout wavelength and diffraction direction of the four superimposed holograms

♯ hologram	Incidence angle ϕ_1	λ_r(nm)	Diffracted angle ϕ_2
1 (10101010)	52°	1549	58.34°
2 (10100101)	52°	1549.8	58.43°
3 (01011010)	52°	1550.6	58.52°
4 (01010101)	52°	1551.4	55.61°

5 Long-Term Reliability of VH-Based Devices

The experimental implementations presented in Sects. 4.1 and 4.2 show the operation of VH-based devices. The volatility of the recorded holograms in a photorefractive material requires them to be fixed in order to design long-lifetime WDM components. In this section we present the experimental results for the fixing and reconstruction of four superimposed holograms read-out at 1550 nm. Holograms have been recorded by the angle-multiplexing technique into a single $LiNbO_3$ crystal. After recording, the thermal post-fixing described in Sect. 3.3 is performed by means of a heating cycle at 137 °C for 20 min. For comparison with the expected fixing efficiency reported in (4), our experimental measurements performed in a $LiNbO_3$ saple doped with 0.015% Fe together with the literature specifications [23] lead to the following expressions:

$$E_{0\text{ph}} = (10^{-15} \div 3 \times 10^{-14}) N_\text{a} \,, \tag{6}$$

$$E_\text{D} = \frac{0.163}{\Lambda} \,, \tag{7}$$

$$E_\text{q} = 9.6 \times 10^{-9} \Lambda \frac{N_\text{a}(N_\text{d} - N_\text{a})}{N_\text{d}} \,. \tag{8}$$

High developing efficiency can be achieved after the developing cycle only in strongly oxidized (large photovoltaic field) weakly (≤ 0.05 mol%) iron-doped crystals because the limiting space-charge field E_q gratelyexceeds the photovoltaic and diffusion fields otherwise. In our case an estimated dopant concentration $N_\text{a} = 2.827 \times 10^{18}\,\text{cm}^{-3}$ and a trap concentration $N_\text{d} = 2.8156 \times 10^{18}\,\text{cm}^{-3}$ lead to

$$\frac{E_\text{D}}{E_\text{q}} \approx 1676 \,, \quad \frac{E_{0\text{ph}}}{E_\text{q}} \approx 791 \tag{9}$$

The relations (9) are in accordance with the conditions concerning the diffusion, saturation and photovoltaic fields required to achieve nearly complete ionic compensation [24]; thus good reconstructed efficiency after fixing can be expected in such a photorefractive crystal. The percentage of the diffraction efficiency which can be reconstructed is very sensitive to a small change of N_a; in particular, the application of an oxidation process resulting in a slight increase of the Fe^{3+} concentration can strongly increase η_fixing.

In our experiment the multiplexed photorefractive gratings were subjected to a heating cycle, followed by a developing stage at room temperature. The diffraction efficiency dynamic behavior of each hologram during recording, fixing and developing processes is shown in Fig. 12. The experiment behavior is in good agreement with the theoretical behavior illustrated in Fig. 6, and the measured reconstructed diffraction efficiency of the fixed holograms

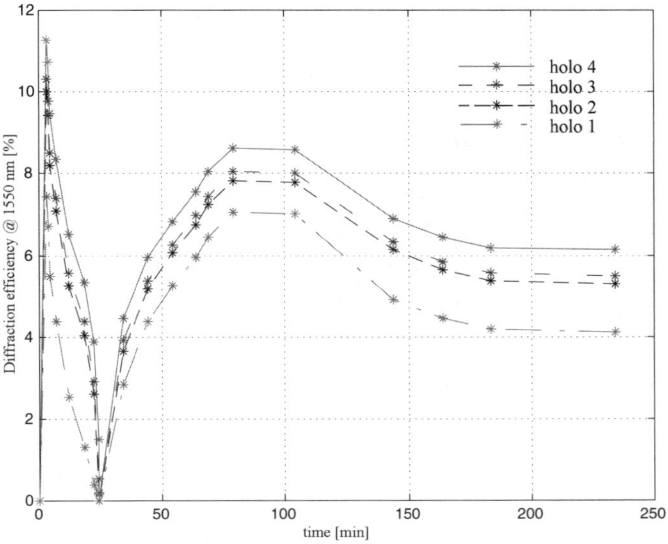

Fig. 12. Dynamic behavior at 1550 nm of four superimposed holograms during the recording, fixing and developing processes

is about 40% of the performed efficiency without fixing, according to (4), (7)–(8).

Substituting the values for our experiment in (3) (H_0 being estimated via the photorefractive method described in [25]), we estimate an approximate lifetime of about 30 years. Recorded holograms for optical communication applications also have to demonstrate a certain robustness to adverse environmental conditions in order to operate in the field for a very long time. We have experimentally tested the robustness to thermal shock of the multiplexed ionic gratings inside a photorefractive $LiNbO_3$:Fe sample by submitting them to long-lasting heating/cooling cycles, with temperature peaks around $+90\,°C/-20\,°C$. The original value of the reconstructed diffraction efficiency, under periodical monitoring, has not shown significant changes. For example, after a 5 h-cycle at 70 °C, the fixed hologram efficiency at 1550 nm displays only a 0.01% reduction, while it is stable after a cooling cycle at −16 °C (Fig. 13).

6 High-Dense WDM Device Design

The experimental results previously presented demonstrate the feasibility of two holographic devices readable by WDM signals in the third communication window spectrum. The main question, already underlined, is how to improve the device performances in terms of the crosstalk level between neighboring holograms and the wavelength selectivity. The most popular solution consists in adopting a different recording configuration, which allows a readout

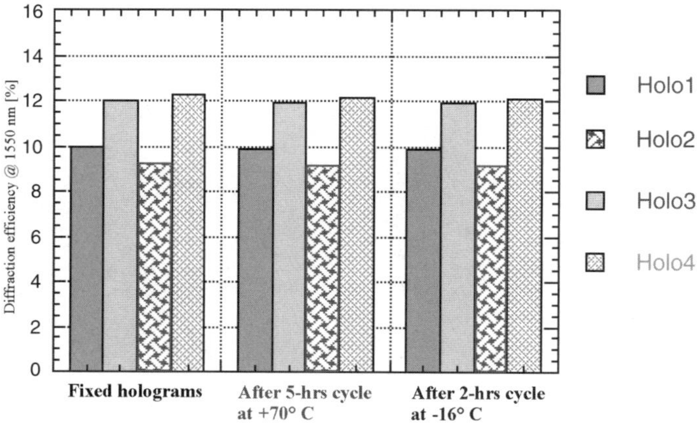

Fig. 13. Performance at 1550 nm of four thermally fixed superimposed holograms, after a heating cycle at 70 °C and a cooling cycle at −16 °C. The insensitivity to termal shock is demonstrated

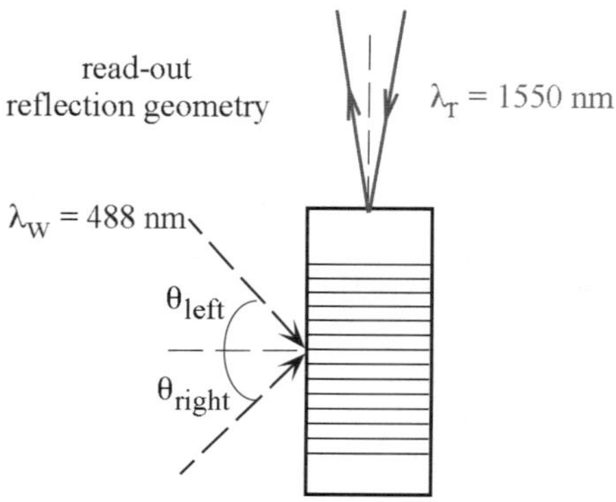

Fig. 14. 1550 nm readout using a reflection geometry

counterpropagating reflection geometry at 1550 nm (Fig. 14); this geometry achieves the highest spectral selectivity.

Such a recording geometry has already been used in the literature to realize NIR narrow-band holographic filters [26,27] demultiplexers [20]. Illumination parallel to the grating wave vector is Bragg-matched for reflection at the vacuum wavelenght λ_{Bragg} given by

$$\lambda_{\mathrm{Bragg}} = \frac{\lambda_{\mathrm{w}} n_{1550}}{sin(\theta)}\,, \tag{10}$$

where n_{1550} is the refractive index at 1550 nm and θ the recording incidence angle. The reflectivity of a lossless uniform amplitude grating in the counter-propagating reflection geometry is [27]:

$$R = \left|\frac{\kappa\,\sinh(-SL)}{(\Delta\,K/2)\sinh(SL) + \mathrm{i}S\cosh(SL)}\right|^2, \tag{11}$$

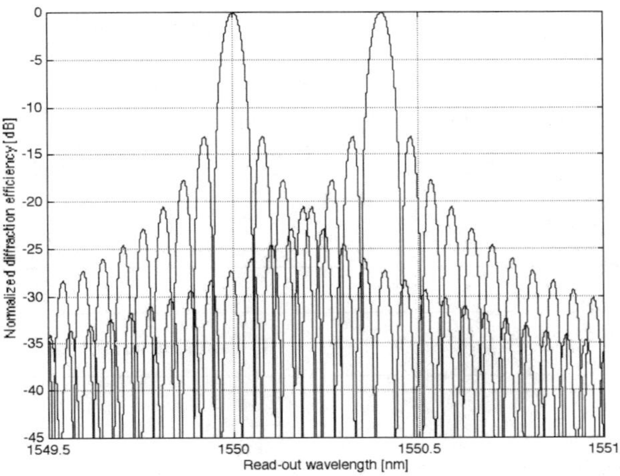

Fig. 15. Normalized diffraction efficiency behavior for two angle multiplexed holograms read in the NIR spectral range by using a counterpropagating geometry

where

$$\kappa = \frac{\pi\Delta\, n}{\lambda_{\mathrm{Bragg}}}\,, \tag{12}$$

$$S = \kappa^2 - (\Delta\, K/2)^2\,, \tag{13}$$

$$\Delta\, K = 4\pi n_{1550}\left(\frac{1}{\lambda_{\mathrm{r}}} - \frac{1}{\lambda_{\mathrm{Bragg}}}\right), \tag{14}$$

with L is the thickness of the grating, λ_{r} is the wavelength of the read-out light, n is the uniform refractive index and $\Delta\, n$ is the amplitude of the refractive-index change. A simulation shows the possibility of separating two channels with 0.4 nm wavelength spacing, showing a crosstalk less than −25 dB (Fig. 15).

The recording and readout parameters employed in the simulation are listed in Table 3. The upper table shows the periods of the recorded gratings, the incidence angle of the recording beams and the assumed photoinduced modulation of the refractive index, while the lower one shows the readout wavelengths, the fixed incidence angle and the direction of the diffracted beams. As the simulation demonstrates, VH-based devices for 0.4 nm spaced WDM signals can be designed with very low crosstalk. The expected performance makes such devices very interesting for future optical networks operating with very HDWDM channels spaced by less than 50 GHz.

Table 3. Parameters for the recording and readout processes of two gratings whose behavior is simulated in Fig. 15

♯ hologram	Λ(nm)	Left-recording angle θ_{left}	Right-recording angle θ_{right}	Δn
1	350.57	44.37°	43.84°	2×10^{-5}
2	350.79	46.9°	41.35°	2×10^{-5}

♯ hologram	λ_r (nm)	Readout angle	Diffraction angle
1	1550	2°	2.4°
2	1550.4	2°	6.2°

7 Conclusions

We have demonstrated the feasibility of VH-based devices, which have applications in optical WDM systems. The implemented devices meet the growing demand for wide optical bandwidth by operating in an all-optical and passive way, without any optoelectronic and electro-optic conversion. Many WDM channels at different wavelengths can be managed both simultaneously and in real time; such possibility is granted by the VH characteristics, which allow us to exploit the inherent parallelism of optics in terms of large storage capacity and data access. We deliberately chose to use a standard photorefractive material as the recording medium, so that the well-established maturity of holographic storage technology could guarantee a very high reliability of the implemented devices, especially in terms of both achievable performance and fixing capability. Operation around 1550 nm, i.e. in the spectral range typical of optical fiber communications, is performed by adopting the two-lambda method in readout, so that the devices written in visible light can be easily used for NIR signal processing. By exploiting innovative VH-based solutions, our final goal is to design new optical devices and components, in order to facilitate the use of optics in all the critical functions of future optical communication networks, from cross-connections to storage.

References

1. Y. N. Denisyuk: Photographic reconstruction of the optical properties of an object in its own scattered radiation field, Sov. Phys. Dokl. **7**, 543–545 (1962)
2. F. H. Mok: Angle-multiplexed storage of 5000 holograms in lithium niobate, Opt. Lett. **18** (11) 915–917 (1993)
3. F. T. S. Yu, S. Wu, A. W. Mayers, S. Rayan: Wavelength-multiplexed reflection-matched spatial filters using $LiNbO_3$, Opt. Commun. **81** (6), 343–347 (1991)

4. G. W. Burr, F. H. Mok, D. Psaltis: Storage of 10000 holograms in LiNbO3:Fe, in *Conf. Dig. Lasers and Electro-Optics*, paper CMB7 (1994)
5. A. Partovi, J. Millerd, E. M. Garmire, M. Ziari, W. H. Steier, S. B. Trivedi, M. B. Klein: Photorefractivity at 1.5 mm in CdTe:V, Appl. Phys. Lett. **57**, 846–848 (1990)
6. M. Annese, P. Boffi, M. Martinelli, 1550 nm photoinduced diffractive grating in CdTe:V, in *Proc. Int. Topical Meeting on Spatial Light Modulators and Integrated Optoelectronic Arrays*, Snowmass, (1999) pp.86–88
7. D. Psaltis, F. Mok, H. S. Li: Nonvolatile storage in photorefractive crystals, Opt. Lett. **19** (8), 210–212 (1994)
8. M. K. Smit, C. van Dam: PHASAR-based WDM devices, principles, design and applications, IEEE J. Sel. Topics Quantum Electron **2**, 236–250 (1996)
9. I. Baumann, J. Seifert, W. Nowak, M. Sauer: Compact all-fiber add–drop multiplexer using fiber Bragg gratings, IEEE Photonics Technol. Lett. **8**, 1331–1333 (1997)
10. A. Pattavina, M. Martinelli, G. Maier, P. Boffi: Techniques and technologies towards all-optical switching, Opt. Netw. Mag. **1**, 75–93 (2000)
11. H. Fujita: Microactuators and micromachines, Proc. IEEE **86**, 1721–1732 (1998)
12. B. Pesach, G. Bartal, E. Refaeli, A. J. Agranat, J. Krupnik, D. Sadot: Free-space optical cross-connect switch by use of electroholography, Appl. Opt. **39**, 746–758 (2000)
13. P. Yeh: *Introduction to photorefractive nonlinear optics*, Wiley Ser. Pure Appl. Opt (Wiley, New York 1993)
14. K. Buse, S. Breer, K. Peithmann, S. Kapphan, E. Kr"atzig: Origin of thermal fixing in photorefractive lithium niobate crystals, Phys. Rev. B **56** 1225–1235 (1997)
15. S. Campbell, X. Yi, P. Ye: Hybrid sparse-wavelength angle-multiplexed optical data storage system, Opt. Lett. **19**, 2161–2163 (1994)
16. X. Zhang, J. Xu, Q. Sun, S. Liu, G. Zhang: Dual-wavelength nonvolatile holographic storage, Opt. Commun **180**, 211–215 (2000)
17. E. Krätzig: Photorefractive effects and photoconductivity in $LiNbO_3$:Fe, Ferroelectrics **21**, 635–636 (1978)
18. E. Chuang, D. Psaltis: Storage of 1000 holograms with use of dual-wavelength method, Appl. Opt. **36**, 8445–8454 (1997)
19. X. An, D. Psaltis, G. W. Burr: Thermal fixing of 10,000 holograms in $LiNbO_3$:Fe, Appl. Opt. **38**, 386–393 (1999)
20. S. Breer, K. Buse: Wavelength division multiplexing with volume phase holograms in photorefractive lithium niobate, Appl. Phys. B **66**, 339–345 (1998)
21. P. Boffi, M. C. Ubaldi, D. Piccinin, C. Frascolla, M. Martinelli: 1550 nm volume holography for optical communication devices, IEEE Photonics Technol. Lett. **12**, 1355–1357 (2000)
22. E. J. Lerner: Advanced applications, Holography: Holographic data storage chips away at barriers, Laser Focus World (Jan. 2000)
23. G. W. Burr, D. Psaltis: Effect of the oxidation state of $LiNbO_3$:Fe on the diffraction efficiency of multiple holograms, Opt. Lett. **21**, 893–895 (1996)
24. A. Yariv, S. S. Orlov, G. A. Rakuljic: Holographic storage dynamics in lithium niobate: theory and experiment, J. Opt. Soc. Am. B **13**, 2513–2523 (1996)

25. R. Müller, L. Arizmendi, M. Carrascosa, J. M. Cabrera: Determination of H concentration in $LiNbO_3$ by photorefractive fixing, Appl. Phys. Lett. **60**, 3212–3214 (1992)
26. G. A. Rakuljic, V. Leyva: Volume holographic narrow-band optical filter, Opt. Lett. **18** (6), 453–455 (1993)
27. V. Leyva, G. A. Rakuljic, B. O'Conner: Narrow bandwidth volume holographic optical filter operating at the Kr transition at 1547.82 nm, Appl. Phys. Lett. **65**, 1079–1081 (1994)

Index

Topics in Applied Physics

69 **Unoccupied Electronic States**
Fundamentals for XANES, EELS, IPS and BIS
By J. C. Fuggle and J. E. Inglesfield (Eds.) 1992. 175 figs. XIV, 359 pages

70 **Dye Lasers: 25 Years**
By M. Stuke (Ed.) 1992. 151 figs. XVI, 247 pages

71 **The Monte Carlo Method in Condensed Matter Physics**
By K. Binder (Ed.) 2nd edn. 1995. 83 figs. XX, 418 pages

72 **Glassy Metals III**
Amorphization Techniques, Catalysis, Electronic and Ionic Structure
By H. Beck and H.-J. Güntherodt (Eds.) 1994. 145 figs. XI, 259 pages

73 **Hydrogen in Metals III**
Properties and Applications
By H. Wipf (Ed.) 1997. 117 figs. XV, 348 pages

74 **Millimeter and Submillimeter Wave Spectroscopy of Solids**
By G. Grüner (Ed.) 1998. 173 figs. XI, 286 pages

75 **Light Scattering in Solids VII**
Christal-Field and Magnetic Excitations
By M. Cardona and G. Güntherodt (Eds.) 1999. 96 figs. X, 310 pages

76 **Light Scattering in Solids VIII**
C60, Semiconductor Surfaces, Coherent Phonons
By M. Cardona and G. Güntherodt (Eds.) 1999. 86 figs. XII, 228 pages

77 **Photomechanics**
By P. K. Rastogi (Ed.) 2000, 314 Figs. XVI, 472 pages

78 **High-Power Diode Lasers**
By R. Diehl (Ed.) 2000, 260 Figs. XIV, 416 pages

79 **Frequency Measurement and Control**
Advanced Techniques and Future Trends
By A. N. Luiten (Ed.) 2001, 169 Figs. XIV, 394 pages

80 **Carbon Nanotubes**
Synthesis, Structure, Properties, and Applications
By M. S. Dresselhaus, G. Dresselhaus, Ph. Avouris (Eds.) 2001, 235 Figs. XVI, 448 pages

81 **Near-Field Optics and Surface Plasmon Polaritons**
By S. Kawata (Ed.) 2001, 136 Figs. X, 210 pages

82 **Optical Properties of Nanostructured Random Media**
By Vladimir M. Shalaev (Ed.) 2002, 185 Figs. XIV, 450 pages

83 **Spin Dynamics in Confined Magnetic Structures I**
By B. Hillebrands and K. Ounadjela (Eds.) 2002, 166 Figs. XVI, 336 pages

84 **Imaing of Complex Media with Acoustic and Seismic Waves**
By M. Fink, W. A. Kuperman, J.-P. Montagner, A. Tourin (Eds.) 2002, 162 Figs. XII, 336 pages

85 **Solid-Liquid Interfaces**
Macroscopic Phenomena – Microscopic Understanding
By K. Wandelt, S. Thurgate (Eds.) 2003, 228 Figs. XVIII, 444 pages

86 **Infrared Holography for Optical Communications**
Techniques, Materials, and Devices
By P. Boffi, D. Piccinin, M. C. Ubaldi (Eds.) 2003, 90 Figs. XII, 182 pages